Evolving to Six Sigma Quality

Yusuf Biviji

Evolving to Six Sigma Quality

A guide on how to integrate six sigma quality
tools with management practices

First Edition

by

Yusuf Biviji

First edition, April 2010

ISBN 1451510985

Printed in the United States of America

Acknowledgements

This book was made possible with the help and support of my loving wife Farzana and wonderful daughters Nabiha and Shazeen. Thanks for all the constructive feedback.

Contents

Introduction

Six Sigma Quality. This phrase is used often but not always easily understood. Several books have proliferated ideas of six sigma quality with some getting into its intricacies, while others just skimming the surface. This book takes on a different approach. It first illustrates the need to use its concepts and tools and then explains them in a simple way. Managers and individual contributors must understand the gains of using these tools before they begin to embrace them. In this book, a story is intertwined with simple management concepts followed by more detailed explanations of various six sigma topics. It is important to know that six sigma is not a loose collection of topics about quality. It is rooted in a common sense approach with defined topics, which if used in a systematic way, will produce results.

This book starts and ends with an introduction. The two introductions however, are very different. One starts the process of understanding six sigma quality tools and the other starts the journey of using them. Using six sigma tools should be the paramount goal of any organization that wants to leap ahead in improving its quality, productivity, and reliability. These three legs of a stool form the foundation for improved customer satisfaction and reduced costs. Together, the two accomplishments impact profits which in turn drive growth. Once these tools are adopted and used, measurable results will follow. Using this book is the first step in that process.

The Way It Is

Managers and individual contributors go through the daily rigor of working with teams and cross-functional groups to show measurable progress. In doing this they sometimes forget their long term objectives and focus on the next big meeting or event. This takes a toll on the success and viability of the individual and also reduces the value they add to the organization. In order to lessen the impact of this day-to-day strife, companies need to invest resources in setting up value added management processes, backed by a data driven decision system. Understanding the concepts of six sigma quality can be a path for organizations and individuals to adopt and use to their advantage. Six sigma tools use cross-disciplinary knowledge that helps build a foundation to manage everyday work. In a sense they form a mosaic of tools, which if used with the right approach, could bring order to a potentially unstructured work environment. Effective management could be achieved with the use of these simple tools, weaved together with a common-sense approach which has to be clear and well defined. It is the lack of clear definition that leads to confusion and inefficiencies in management practices.

The word 'mosaic' conjures up images of designs and patterns, laid out in vivid colors. Managers of today have to use a similar approach, which requires them to be skilled in multiple areas, know how to manage different situations, and learn the patterns of the corporate world. The net output of their effort should be a work of art, in which defined visions of success become a reality, looking like a mosaic, and not a hodgepodge of uncoordinated actions. They have to create, sustain, and improve their work practices to achieve desired objectives. The bar is constantly being set higher. Results need to be concrete and measurable, which will become a reality once the practices outlined in this book are implemented. Results should also be more than just meeting a number. They should include setting up a system in which sustainable gains are made by understanding the intricacies of the job on hand.

The management practices outlined in this book are narrated in a story. In this story, the central character Jeff works hard without realizing the importance of how to streamline management practices and use data driven tools to make his work more effective and rewarding. It is said that 'the teacher arrives when the student is ready'. When Jeff is ready to accept and learn, he finds that his boss Dean is a willing teacher, and not just a dissatisfied executive. In the end, Jeff begins to spend less time on administrative work, and more on developing new and improved processes. From a time study perspective, the percentage of time spent on 'non-value added' activities should be minimized. Correspondingly, time spent on 'value added activities' such as developing ideas, designing processes, and innovating should be maximized.

This book is a step-by-step guide on how we can restructure the way we work, while laying the foundations of valuable six sigma tools. Each of the steps will produce a certain defined outcome. This outcome will put us closer to a 'desired state'. It is the pursuit of this desired state, which should drive the change process. After each chapter, take a moment and visualize the desired state for your circumstances, before attempting to change anything. If the desired state is appealing to you, then it should be reason enough to take action and move towards it. It should also provide motivation to learn the tools of six sigma quality and use them as a catalyst for improvement.

The Cast of Characters

Here is part of the organization which is referenced in the story to follow. The empty blocks represent others who are not a part of this story.

Organization Structure:

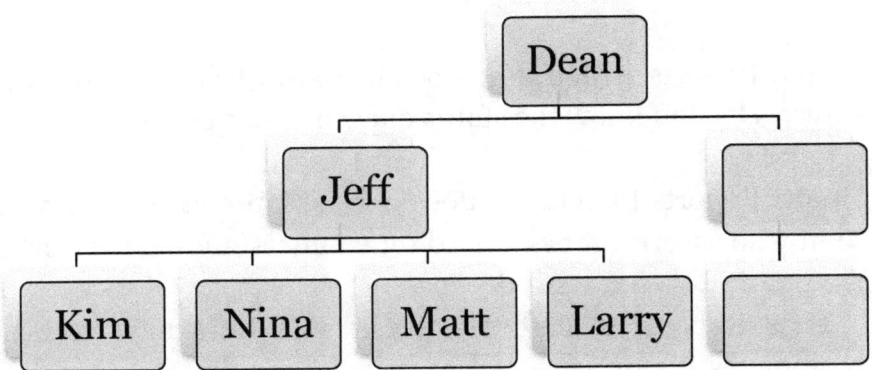

Jeff: Our central character. The one who learns it's not too late to begin. He risks his job to invest time in new learning and become a better manager.

Dean: The executive, and Jeff's boss. He is like any executive you'll find except that he goes out of his way to guide Jeff on how to improve fundamentally as a manger.

Kim: Reports to Jeff. She knows the ins and outs of her job, but does not summarize her findings in a succinct way for Jeff to understand the results.

Nina: Reports to Jeff. She does her work well, but does not always let Jeff know the status of each of her projects.

Matt: Reports to Jeff. He does not communicate much with Jeff who as a result has a hard time understanding his work.

Larry: Reports to Jeff. He doesn't let Jeff know how he completes his projects. All he says is "it's done" and moves on to the next project.

Chapter 1: The Morning Rush

Jeff rushed out of his car and headed straight for the building entrance. It was just past 7:00 in the morning, and he wanted to reach work ahead of his boss, Dean. Jeff worked long hours to make sure that he got his job done. His team of four people worked on various projects to improve the company's operational performance. Although their work was supposed to show the value they add to the business, it did not always turn out that way, which is exactly what happened that morning.

When Jeff got to his desk, he saw that Dean was already in the office. Jeff usually tried to get in a few minutes before him. A comment, such as the one he made the other day, "working hard, Jeff," was music to his ears. Anyway, Dean beat him to work that day, but Jeff still wanted to make sure Dean knew he had arrived. After stopping by his desk to drop off his bag, he walked by Dean's office. From the corner of his eyes he could see that one of his employees, Matt, was sitting at the desk, explaining something that sounded like "I wanted to make sure that you were aware I have put in all this effort." "Frog Jumping!" Jeff thought as he went to the break room and made a cup of coffee. He did not bother to add any cream and sugar, as he wanted to get back and catch up on Matt's conversation. Mustering some courage, he decided to step into Dean's office.

To his surprise, he saw that Matt had just left and Dean was on the phone. "Not a great start to the day," thought Jeff. Waving to Dean, he stepped out of his office. It seemed that

Dean had not noticed him, since he kept talking on the phone, without acknowledging his entry or exit.

Back at his desk, Jeff booted up his computer, and wondered what that conversation had been about. He decided to meet Matt and ask him. As he approached his cube, he saw him walking away to the cafeteria, presumably to get some breakfast. "Oh well, I'll get to the bottom of this conversation later. Let me put my brain back into gear and catch up from where I had left off yesterday," he thought.

The day whizzed by and Jeff never got a chance to follow up on Matt's morning conversation. First, Dean walked up to him in a hurry, and wanted a status update on the 'Situational Analysis' project. Since Nina was working on it, Jeff had to quite embarrassingly admit that he did not know where it stood, and would get with her to find out if it was done. During that time, he received a voice mail from Larry, saying that he had caught the flu bug, and would not come in for the day. He realized that Larry was supposed to present to the board of directors that morning on his findings on the 'Madison' project.

Jeff's heart skipped a beat. Frantically he looked around trying to find a hard copy of the presentation. When he did not find one, he called Larry back and asked him where the files were on the shared drive. Quickly he printed a color copy. He had a few more minutes before the presentation and wondered if he should find out more from Nina about her project or call Larry back to determine what he had to say in the meeting.

He decided to put Nina's work on the back burner, and called Larry instead. This time, the phone kept ringing. When the answering machine kicked in, Jeff left a frantic message. Larry did not call back, and Jeff, feeling a bit dizzy, walked up to the meeting room to present something he didn't know much about.

The presentation was a haze. Jeff remembered that Dean was asking him some difficult questions. "Whose side is he on?" he wondered later. "Instead of coming to my defense, he treats me like an outsider." Jeff could not figure Dean out, although he had worked for him for almost a year now.

Jeff's biggest challenge was to make Dean understand that it was not easy being a middle manager, especially since he worked on so many projects simultaneously. "It is different managing managers," he convinced himself that evening. "Dean doesn't realize how easy he has it with conscientious people like me reporting to him." With the kind of team that Jeff had, it was difficult. His employees were carefree and in his opinion, didn't take their work too seriously. They did not seem to have the structure and routine that he loved. This was making Jeff's work a daily struggle and he did not quite know the way out.

Chapter 2: An Early Start

The next morning, Jeff got to work at 6:00 a.m., to make sure he was ahead of Dean. He had had enough of the run around and was tired of fighting fires day after day. "I have to do something that will make me leverage the benefit of having a team work for me," he thought. "Having people work for you should be an asset, not a liability. From the way things are working out for me, they are a huge liability. From not giving me the status of current projects, to not taking important presentations seriously, to leap frogging the immediate supervisor, all this is causing me to de-focus from how I am going to improve our department's value, to just saving my skin. All this, in just one day; I have to do something about it."

Once Jeff got to his desk, he grabbed a pen and started to jot down what his team should do to add more value in a concerted manner. "There should be basic guidelines which describe our attributes. These attributes should define the level of expectations for all." Jeff's mind was racing now and he was thinking faster than he could write. Soon he came up with one page of what he thought should be the baseline of expectations. It went like this:

Set expectations for the team: Without correctly setting expectations for the team, it is unrealistic to think they will perform at required levels. Therefore, the first step is to set expectations and then strive to have the team achieve them. Your goal is to form a 'star' team, which meets and exceeds these expectations. First make sure that you personally meet or exceed them yourself then communicate the same to the team. Some of the attributes that you may find useful include:

(a) **Demonstrate personal integrity:** Be honest in what you do and what you say. Personal integrity is the base on which other attributes are built. Without this, nothing else really matters. Be true to yourself, and to the organization regarding hours worked, correctly reporting data, ethical use of company resources, etc.

(b) **Meet basic job requirements:** Maintain good attendance, arrive to work on time, put in the required hours per week, attend meetings on time, and perform basic tasks that are needed to 'get the job done'. This constitutes the bread and butter part of the job and must be met on a continual basis.

(c) **Have the ability to learn quickly:** Be adaptable to learn new software, hardware, tools, and processes. Be able to adjust your mindset to accept new business needs and goals. Integrate learning into daily job functions, so as to constantly improve performance.

(d) **Have the ability to work independently:** Stay customer-focused, and deliver what was promised within the required timeframe. Make sure that you do not need to be micromanaged and all job requirements are met on a daily, weekly, and monthly basis. Effectively communicate when targets are not going to be met, so that there are no surprises in the end.

(e) **Work well in a team:** Consider and respect others' points of view, and enhance the team's output by contributing positively. Provide ideas to improve the team's performance. Participate in team activities.

(f) **Possess good communication skills:** Possess effective verbal, written, and active listening skills. Check for understanding, and follow up with others to ensure that work is progressing as designed. Use your communication skills to drive performance from peers. Highlight your work through proper channels to gain credit for your efforts.

(g) **Make quality and productivity the hallmark of your mission:** Constantly look for opportunities to improve processes and recognize issues quickly when they occur. This mindset will enhance the organization's quality and productivity. Do not wait for a 'tell'. Have the skills and ability to identify these opportunities.

Desired State

Your team has the attributes listed above. Those who don't, thoroughly understand them, and are actively pursuing to attain them. Those who did not have the capability to accomplish them have moved to a job function that better suited them.

Once Jeff was done writing, he read through what he had written, and felt pretty good about it. He hadn't done anything concrete yet, but this was at least a start. He believed that quality and productivity needed to be nurtured as a driving goal within an organization. The area of quality stemmed from years of quality leadership and research which had led to the richness of this field. There were many leaders in the quality field. Jeff researched some of them, to gain insights into their thinking. He realized that their contributions have meant a great deal and form the foundation of the quality thought process of today. "When quality and productivity fit like a hand in a glove," he thought, "then business processes start improving in a sustained manner." Jeff reviewed the works of some of the quality leaders who are listed on the following pages.

Quality Leadership

Quality leadership has had a long history over time. Through the years, there have been quality leaders who expressed thoughts that have changed people's view of quality. From being a 'necessary evil' or even something that could be ignored in the past, to the more recent use of quality as a strategic advantage, quality has come a long way. There have been many contributors to the quality field, and the input of some of these prominent thought leaders are described below:

1. Edward Deming
2. Joseph Juran
3. Philip Crosby
4. Kouru Ishikawa
5. Genichi Taguchi

Each of these leaders had their own views on quality, which in some respects were not very different from the other leaders, but provided a new dimension to the body of knowledge for quality. Each of them proposed some aspect of quality which evolved the range of its ken.

Dr Edward Deming wrote the book 'Out of Crisis'. Deming was trained as a statistician and helped during World War II to improve the quality of war materials. He was invited to Japan at the end of the war by Japanese industrial leaders and engineers, where he helped them build quality systems. He developed his quality principles, also known as Deming's 14 points and propagated their use. These points have been widely read and used all over the world. They are:

1. Constancy of purpose
Create constancy of purpose for continual improvement of products and services to society, allocating resources to provide for long range needs rather than short term profitability. Plan to become competitive to stay in business and provide jobs.

2. The new philosophy
Adopt the new philosophy. Since we are in a new economic age that was created in Japan, we can no longer live with commonly accepted levels of delays, mistakes, defective materials, and defective workmanship.

3. Cease dependence on mass inspection
Eliminate the need for mass inspection as the way of life to achieve quality by building it into the product upfront.

4. End lowest tender contracts
End the practice of awarding business solely on the basis of price tag. Instead, require meaningful measures of quality along with price. Reduce the number of suppliers for the same item by eliminating those that do not qualify with

statistical and other evidence of quality. The aim is to minimize total cost, not merely initial cost, by minimizing variation.

5. Improve every process
Improve constantly and forever every process for planning, production, and service. Search continually for problems in order to improve every activity in the company to enhance quality and productivity, and decrease costs. It is management's job to work on the system and foster innovation.

6. Institute training on the job
Institute modern methods of training on the job for all, to better use every employee. New skills are required to keep up with changes in materials, methods, products and service design.

7. Institute leadership
Adopt and institute leadership aimed at helping people do a better job. The responsibility of managers and supervisors must be changed from sheer numbers to quality. Improving quality will automatically improve productivity. Management must ensure that immediate action is taken on reports of defects, maintenance requirements, poor tools, fuzzy operational definitions, and all conditions detrimental to quality.

8. Drive out fear
Encourage effective two way communication and other means to drive out fear throughout the organization so that

everybody may work effectively and productively for the company.

9. Break down barriers
Break down barriers between departments and staff areas. Also people in varied areas such as Leasing, Maintenance, or Administration must work in teams to tackle problems that may be encountered with products or services.

10. Eliminate exhortations
Eliminate the use of slogans, posters and exhortations for the work force, demanding zero defects and new levels of productivity, without providing methods to do so. Such exhortations only create adversarial relationships; the main causes of low quality and productivity belong to the system, and thus lie beyond the power of the work force.

11. Eliminate arbitrary numerical targets
Eliminate work standards that prescribe quotas for the work force and numerical goals for management. Substitute this with helpful leadership to achieve continual improvement of quality and productivity.

12. Permit pride of workmanship
Remove barriers that rob hourly workers and people in management of their right to 'pride of workmanship'. This includes abolition of the annual merit rating and of management by objectives. The responsibility of managers, supervisors and foremen must be changed from sheer numbers to quality.

13. Encourage education

Institute a vigorous program of education and encourage self improvement for everyone. An organization needs not only good people, but also those who are improving with education. Advances in competitive position will have their roots in knowledge.

14. Top management commitment and action

Clearly define top management's permanent commitment to ever improving quality and productivity and their obligation to implement these principles. It is not enough for top management to just commit themselves to quality and productivity. They must know exactly what it is they are committed to. Create a structure to push on the preceding 13 points, and take action to accomplish the transformation. Words of support are not enough: action is required!

Dr. Joseph Juran wrote the 'Quality Control Handbook'. Juran developed the Quality trilogy, which can be divided into three stages:
1. Quality Control
2. Quality Improvement
3. Quality Planning

Quality Control

The quality control part of Juran's trilogy identifies and removes special causes to maintain process stability. This is the traditional reactive view which is not focused on prevention, but does attempt to improve processes. In this traditional view, control charts are used reactively to monitor process behavior.

Quality Improvement

The quality improvement part of the trilogy refers to reducing common causes in a process to create a more stable process operating at an improved quality level. This proactive approach does not accept the constant generation of problems, and uses prevention as a mechanism to reduce special causes from occurring.

Quality Planning

The quality planning element deals with advanced quality planning to design low variability processes in which common and special causes are minimized.

Dr. Philip Crosby wrote the books 'Quality is Free' and 'Quality Without Tears'. His books were easy to read, and thought provoking. He introduced quality concepts which can be summarized in the four absolutes for quality management:

1. Quality is defined as 'conformance to specifications'
2. The system for improving quality is prevention, not detection
3. The performance standard for quality must be 'zero defects'
4. Quality should be measured in terms of 'the price of nonconformance'

Each of the above points has great depth when reviewed in detail. The first point talks about defining quality as 'conformance to specifications'. This definition of quality is widely accepted today, and the credit goes to Crosby. It links customer expectations, to product or service specifications. The second point talks about using prevention, and not inspection to improve quality. This is similar to what Deming and Juran state and is very powerful when applied in the real world. The third point is about the concept of 'zero defects'. Deming does not approve of such slogans, but if viewed purely as a goal in which there is no tolerance for failures, it becomes a driving force to understand gaps between current performance and the ideal state. Once this gap is understood, steps can be taken to narrow it. The fourth point talks about the actual cost of quality, which if measured as the price of nonconformance, would include a host of items that are not traditionally counted in the cost structure.

Kouru Ishikawa invented the cause and effect diagram which is also called the 'Ishikawa' or 'fishbone' diagram. Using his quality tool, and also other quality tools, he made significant advancements in quality improvement efforts. With the use of this new diagram, the user can see all possible causes of a result, and find the root causes of existing problems. He expanded on the four step 'plan-do-check-act' into the following six:

1. Determine goals and targets
2. Determine methods of reaching goals
3. Engage in education and training
4. Implement work
5. Check the effects of implementation
6. Take appropriate action

Genichi Taguchi first proposed the concepts of 'orthogonal arrays', 'robustness in design', and 'loss function'. These concepts helped add new dimensions of thought to the quality field. They are briefly described below:

Orthogonal Arrays: In a process, outside factors or 'noise' cause deviations from the mean. Taguchi devised a way to use a tool called 'orthogonal arrays' to isolate noise factors in a cost effective manner.

Robustness: Taguchi referred to robustness as the ability of a process or product to work as intended regardless of uncontrollable outside influences. He was pivotal in developing many companies' products and processes which performed uniformly regardless of uncontrollable forces.

The Loss Function: Taguchi devised an equation which showed the relationship in the decline of a customer's perceived value of a product to the decline in its quality.

All the above quality leaders, in addition to many others, have contributed to the quality field and made it richer by adding new perspectives to its various aspects. They have broadened its horizon and incorporated new content which built upon or even replaced existing material. Quality is still evolving, with new and innovative approaches to improve it. This will make it richer in the future. Closely related to the concept of quality is that of productivity. In an ideal scenario, productivity improvements should lead to better quality and cost savings.

Productivity

Productivity is also called efficiency. Simply put, productivity is output divided by input. This means that as you increase output, with the same or less input, you increase productivity. Similarly, if output is reduced, and input is increased, productivity goes down. This definition can be used in various scenarios, such as equipment performance, process performance, or even people performance. Improving productivity increases the profitability of a company, so it is something that organizations strive to do. However, there is also a cost involved in making productivity gains that needs to be balanced with output gains. Sometimes, productivity gains are realized by reducing inspections or other controls in the system to drive down input costs. This could result in reduced quality. The best productivity gains are those which result in a corresponding increase in the quality of output.

Productivity tools

There are several tools that can be used to measure and improve productivity. Some of them are:

1. Time and motion studies
2. Process re-engineering
3. Value engineering
4. SMED (Single Minute Exchange of Dies)
5. Poke-yoke (also known as error proofing)

Based on the particular scenario, one or more of these tools may be used to improve productivity of a process.

Time and motion studies: This is a simple tool which breaks down complex processes into smaller elements, and categorizes each element as being either value added or otherwise. An important objective of this study is to determine the sequence of movements done by the person performing repetitive tasks. The goal is to eliminate redundant or wasteful motion. Based on these studies, the 'standard time' for particular tasks can be calculated and used to estimate the amount of work that should be completed in a day or shift.

Process re-engineering: This tool takes the approach that redesign and reorganization of a process by 'wiping the slate clean' can lead to lower costs and better quality. It encourages step function changes rather than incremental improvements. This kind of change is usually accomplished with 'out of the box' thinking and use of information technology as a key enabler.

Value Engineering: Value Engineering was developed at General Electric Corp. It is an analysis of the functions of a product, aimed at improving its performance, quality, reliability, safety, and life cycle costs. By subjecting a product through the value engineering process, a more robust product can be created.

SMED: This term literally means changing dies in a machine in less than 10 minutes, hence the term 'single minute exchange of dies'. In a broader context, it stands for designing a process in which all preparatory work is done, and only the most critical tasks are undertaken while the system is down, so as to minimize system downtime, thus improving productivity. This kind of thinking can save an organization from having machines, systems, or any other operation down for an extended period of time, by ensuring that everything that could be done to minimize downtime is complete prior to the actual event.

Poke-yoke: This is the Japanese term for error proofing. Mistakes usually happen if there is a way to make them. If the opportunity to make errors is removed, and there is only one way to do a task, then errors will be eliminated. When processes, designs, or user interfaces are error proofed, quality of the output is much improved. Special error proofing efforts should be undertaken within organizations to prevent the occurrence of problems. Implementation of Poke-yoke improves both productivity and quality.

Productivity from an economic standpoint

At a broader level, improvements in productivity drive increases in the standard of living. As we get more efficient as a society, our outputs increase with the same inputs, thus improving the overall quality of life. At the corporate level, increased productivity means reduced costs. Productivity is a two-edged sword when it comes to quality. Eliminating parts of a process which contribute to quality would improve productivity, but reduce quality. Eliminating non-value add parts of a process, re-engineering a process, or innovating a different solution, could actually improve quality, while simultaneously improving productivity. It is this latter method which causes a permanent shift in productivity levels, and should be pursued by organizations. Automation also drives improvements in productivity and its implementation should be balanced with cost and quality.

Chapter 3: Confession

Jeff was amazed at how quickly he had come up with a list of expectations for himself and his team. "Pretty basic," he thought. "How did I work and manage all this time, without defining these expectations?" His next step was to socialize this list. He wanted to be firm but gentle in informing his team of the expectations he had from them. The last thing he wanted was to sound haughty about it. He thought that the next team meeting would be a good time to talk about this, and solicit feedback on it. Although what he had written down was general in nature, the discussion he intended to have with his team would focus on specifics. The important thing was that he was now quite sure of himself, so feedback would help him with the nuances of his writings and not the general intent.

Feeling good about what he had come up with did not last too long. The phone rang, and it was Larry again. His flu was still bad, and he was very sick. "Take rest and don't come in today, lest you spread those germs to all of us," said Jeff in an authoritative voice. After the call Jeff thought "Wow, how am I going to get Larry's work done now? It's his second day out, the presentation did not go all that well, and updates to the open items fielded during the meeting are due tomorrow." Jeff started wringing his palms, wondering what to do next. He wasn't sure where the data had come from, how things were calculated, and the basis for the recommendations. He wasn't good at using the tools that Larry used on a daily basis, and didn't know for sure who

Larry had talked to for getting the information. All Larry had told him two days ago was that the project was complete, and Jeff was glad to hear that. Jeff believed in giving his employees their own space, and did not like to micro-manage them.

Jeff gave the situation some thought. Honesty and integrity meant a lot to him. He decided to come clean with Dean, and talk to him about it. Luckily, he saw Dean walking round the corner, and waved to him. Dean ushered Jeff into his office. Jeff started talking about Larry's work, and how he wasn't sure about its details or how it was done. Dean interrupted him and asked "Wait a minute, are you telling me that you do not have any concept of how this work was done? Let me tell you this Jeff, if you were to take a few unexpected days off, I would be able to cover for you. The reason is that I know the basics of not only your job, but also the job of your employees. Granted, I may not use the tools they have on a regular basis, but I do understand them. I would be rusty at using them, but would surely be able to tell someone with the right skills, on what to do. It is important for you to learn your employees' work. I do realize that you are busy with other day-to-day tasks, but it is well worth the investment of your time to learn your team's work. Let me take some time, to write this down for you, so it will help you reinforce what I just said." Dean proceeded to type his words of wisdom. He continued to talk and type, while Jeff waited listening to his council and the clicking sound of the keyboard. He then handed Jeff a sheet of printed paper which read as follows:

Understand the basics of the team's work: These are skills needed to get the job done by the team. As a team leader, you need to understand what each person does in his or her job. To do this, you need to be well versed in the activities they do. By learning the skills your team uses in their day-to-day activities, you will be able to 'command' their respect, rather than having to 'demand' it. This will also protect you in case someone has to take a long leave. You will have the knowledge and ability to train a new person at that point, and cut your losses.

Dean then went on to say, "I understand that you may not want to be a master in each and every aspect of your team's work, but do not fall into the trap that a manager need not know the technical aspects of the team's work. Only by understanding the details can you lead efforts of thinking outside the box to get things done better and faster. At the very least, I would want you to know enough to be dangerous."

Desired State
As a team leader, you understand the basics of each team member's work. You know what pitfalls they face and how to resolve them. In essence, you could easily collaborate with and direct another subject matter expert on how to proceed with any of your team member's job in their absence.

That was quiet a bold statement by Dean. Jeff read his writings, and although hated to admit it, he did agree with it. He knew that he had to invest time on the tasks done by his team. He also knew that at a higher level, he needed to understand more about quality and reliability. In order to know the technical aspects of any work, it was necessary to understand the concept of reliability. It was important to learn about data analysis and the foundations of quality. He decided to read up on content related to reliability first, since it was often neglected but was in fact, the foundation behind a lot of the work they did.

Reliability

Reliability is an area which covers the longer term aspects of quality. It is the ability of a part or a system to continue to perform normally over a specified amount of time. Reliability deals with failure rates of parts or systems. When failure patterns of entire populations of parts or systems are observed, we see a trend in how they fail. This pattern is captured in the form of a curve, which based on its shape, is known as the 'bathtub curve'.

The Bathtub Curve

The life of a population of parts can be divided into three separate phases, as shown in a 'bathtub curve'. This curve shows a typical failure pattern for parts, from the start to the end of their life. The first phase is the downward portion of the line, which signifies a decreasing failure rate. This corresponds to the early life of parts. During this time, defective parts fail, leaving behind a more robust population. The next phase is the flat portion of the line, which corresponds to the normal lifespan of parts. During this time, failures occur randomly and the failure rate is constant. Hence the slope of the line is also constant. The third phase starts when the slope begins to increase until it reaches the end of the graph. This happens when parts become old and fail at higher rates.

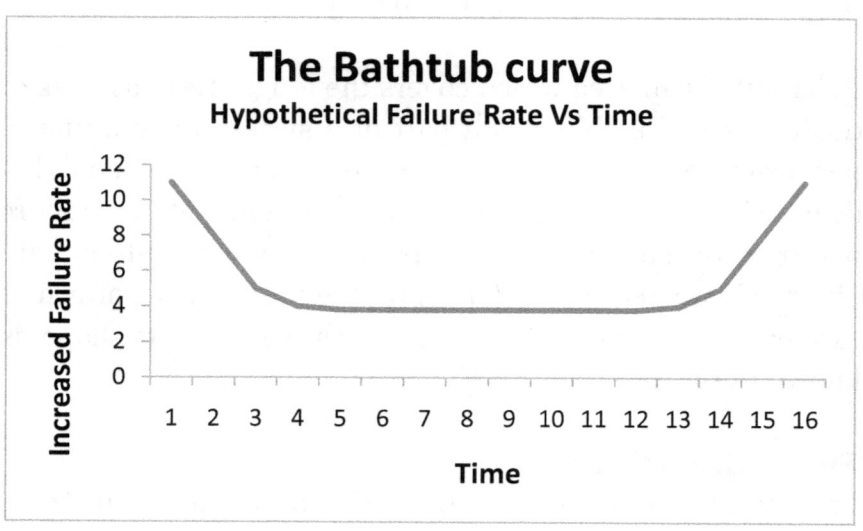

Reliability Calculation

Reliability is defined as the mean time between failures (MTBF) for repairable items and mean time to failure (MTTF) for non-repairable items. This means that reliability is the average length of time a part will work in a normal state.

Mean Time Between Failure (MTBF)

As described above, MTBF is a measure of reliability for repairable items. It is the number of hours, days, months before a part or system fails. For example, if a part has a failure rate of 5 failures per 1000 hours, the MTBF is:

MTBF = Total Time/total number of failures
                = (1,000 hours) / (5 failures) = 200 hours

Hence, MTBF is the inverse of the failure rate. As a caveat, this applies when the failure rate is constant.

Mean Time to Failure (MTTF)

MTTF is used in reliability calculations for non-repairable items. It is the average time until the first failure occurs. Just as for MTBF, MTTF is also the inverse of the failure rate, and is applicable when the failure rate is constant.

Although MTBF should be used only for repairable items and MTTF for non-repairable items, often MTBF is used for both repairable and non-repairable items.

Failure Rate

If MTBF is known, the failure rate is the inverse of MTBF.

Weibull Analysis

Weibull analysis is used as a method to determine where a population of products is on the bathtub curve. It is extensively used in several reliability analyses.

Accelerated Life Testing

Accelerated life testing shortens the life of a product or quickens its degradation by using high stress tests. The goal is to obtain performance data ahead of time, which simulates field conditions. This kind of testing gives an estimate of the reliability of a product before it is launched. The key is that tests should be accurate enough to match field conditions, and user behavior. Often, lab testing may not reflect real use patterns, so results may not be very accurate. If tests are close to reality, then these tests provide early insights into how products will perform in the field.

Failure Mode and Effects Analysis (FMEA)

FMEAs are used to predict potential failure modes, so that they can be prevented. There are two kinds of FMEAs; the design FMEA, and the process FMEA. FMEAs use the risk priority number, or RPN to rank risks associated with each potential failure. The RPN is a straight multiplication of the occurrence score, severity score, and detection score. Prior to starting a FMEA exercise, the team should define and agree on the scoring method to assign scores from 1 through 10 for the occurrence, severity, and detection. Then, for each failure mode, all three scores are multiplied to get the RPN.

Chapter 4: Revelation

Jeff had mixed feelings about his encounter with Dean. On one hand, his pride was hurt with all of Dean's preaching. On the other hand, he felt that he gained a lot from what Dean had to say. "Why did it take him a year to tell me something so basic," he wondered. But then, it had taken Jeff the same amount of time to confess that he did not have any concept of the work of one of his own team members. Jeff read and re-read what Dean had written for him. He now understood why it was important for him to know how Larry did his job. It wouldn't amount to micromanagement, he realized. "All I need to know is how Larry gets the job done. Then, when I need a status update, I would know exactly the kind of effort that went into doing the work. Also, I would need to know it well to improve on it." This made good sense to Jeff, but he still felt that he was missing out on something. It was like a piece of the puzzle that was not coming together.

Jeff started to think aloud now. "How do I know if Larry really knows his job well? He says he is the expert and I have assumed that all along, but I haven't validated it. If there isn't any documentation of his work elements, how would I understand his expertise?" Jeff knew that he was on the verge of a new revelation. He felt thoughts flowing freely, and did not want to stop them. He set aside his worry about next day's presentation, got a marker, and started writing on his board:

Document all elements of the work done by the team. No part of it should be 'black boxed'. This comes down to understanding what each team member does and if there is adequate documentation of their work. To document work, review all elements of the job with each team member.

Jeff knew if he wrote down what each person on his team should know, it would shed light on a lot of areas that they would need to improve on. It was going to be a difficult task, but not doing it would be more difficult in the long run. "This last year has been crazy, and I'm not going to let the next year be the same," he resolved. He started to write down a job requirements sheet for Larry. "This is what I would consider a baseline document," commented Dean later. "It needs to be more specific, though. When you say 'requires knowledge of Excel', you need to specify what knowledge is required. Is it creating macros, data tables, pivot tables, or all of them? Identify skills you need your team to know, and then list them down. This will allow you to objectively evaluate your existing and new team members. It will also help you determine how to develop their abilities with the passage of time."

"That summed it up pretty well," thought Jeff. "This is indeed a foundation laying process towards good, progressive management." Jeff decided to develop a baseline document for each job function and list the skills and knowledge required to perform the job. As he started writing down the requirements, he realized that he wasn't sure about how much training some of the skills needed. It wasn't a one person task to create these documents. He would have to share them with his peers and team members, and iteratively develop them.

Desired State
All team members know how to perform their jobs effectively. Their job elements are well understood and documented. They are the 'experts' for the position they hold.

Jeff liked the thought of the desired state and wanted to make sure that this was not a one-time activity. He continued writing to build upon what he had so far:

Once you understand and document all elements of the team's work, it is time to challenge it. Ask questions such as 'Is work done by the team in line with the overall goals of the organization?' 'Is any of this work obsolete?' These questions will help challenge the current work. Also, become an advocate of the customer and ask if all this effort helps them. Understand who they are, what they want, and the value your work will add in meeting their needs. All this could be encapsulated in the term customer focus. Customer focus is the methodology of doing work, so that it adds value to your product or service from the customers' point of view. The voice of the customer needs to be heard and spread through all processes of the business. This will provide a good frame of reference while deciding if the tasks on hand are worthy of being pursued.

The voice of the customer should be gathered through internal and external surveys. These surveys will give you direct insight into what they are thinking about. Since the customer is the last person in the food chain for delivering the service or product, they should be asked about their opinion. This should be done as part of completing the sale, or providing the service. If this step is added to the process, feedback from the customer will become a part of everyone's thinking.

Another approach to ascertaining customer satisfaction is via monitoring. Monitoring is a broad term, and could mean a lot of things. One of its objectives is to track customer interactions, and determine if they seemed to be satisfied. Another approach is to analytically deep-dive into buying patterns of customers to understand what makes them like the service or product, and what drives sales.

Both surveys and monitoring help in developing customer focused metrics. These metrics look at things from the customer's point of view, and center around the end user experience.

Jeff wanted to gain more insight into the areas of customer focus and metrics which would improve product and service satisfaction. He knew that incorporating this into the team's objectives would make them work on important things. He decided to read up on these two topics, and glean what he could from them.

Customer Focus

Although there are several definitions for 'Quality', all of them either directly or indirectly think of the customer. Traditionally, we think of quality as 'conformance to specifications', or 'reduction in variation'. These definitions hold true for both, products and services. By delivering a more consistent product or service, with reduced variation, the customer gets something that's close to what they had been promised. Other definitions of quality are bolder, directly linking performance to customer satisfaction, saying that quality is providing the customer what they expect, or quality is meeting and exceeding customer expectations. For all definitions of quality, the customer is always central in thought. The question though, is how would one know if customers' needs are truly being met? There are several ways in which the 'voice of the customer' (VOC) can be heard. Some of them are:

1. Surveys
2. Monitoring
3. Focus groups
4. Complaints or warranty tracking
5. Use of customer focused metrics
6. Understanding customer behavior patterns

Surveys
External quality data may be obtained from surveys, which ask customers a series of questions to judge their experience with companies, products, or services. This data is important, as it gives insight into what customers think and

allows the evaluation of processes based on their feedback. In that sense, this data helps meet an important definition of quality, which is reduction in variation of the service or product. Surveys may also be conducted internally, to determine improvement opportunities for a group, department, or person. Although surveys do represent a very important channel of information, there are two points to keep in mind about them:

1. Survey information is 'after the fact', and some damage may already be done to the brand/image if the organization waits for surveys to learn about what went wrong.
2. Survey data should be quickly acted upon for any chance of recovery in customer perceptions.
3. Survey data is qualitative in nature, and should be viewed accordingly.

Monitoring

Quality data may be obtained from one or more individuals or systems called 'quality monitors', who collect data from customer interactions. They could also track several other issues of interest, based on service type and business requirements. This is usually a technical evaluation of quality, and tends to satisfy another important definition, which is 'conformance to specifications'. Quality monitoring can also be automated in some instances, where pre-set triggers set off an alarm to concerned parties that certain thresholds have been exceeded with some dimension of customer interactions. For example, if requests for purchases over the web are queued up beyond a certain point, then it is an indication that customer satisfaction will be impacted.

Quality monitoring informs us of problems before any survey data comes in, so it is a faster means of identifying and correcting problems.

Focus Groups

A focus group is a great tool to gather detailed information on a product's or service's features, benefits, and issues, as seen through the eyes of the average customer. This technique involves selecting customer groups (either randomly, or based on certain patterns), to ask them questions about what they like, or do not like about the product or service. One could get innovative with focus groups, getting into areas such as deep seated customer perceptions, and how current or potential customers would react to new offerings. This is a qualitative method of gaining insights, directly from customers. It is important to ensure that the feedback received from this source is not biased in any way.

Complaints or warranty tracking

Another great source of customer data is to look at complaints and returns. This source gives insight into extreme cases, or failures which occur along the way in delivering products or services. Warranty is a hard source of data, in which we can count the number of returns for defective parts. Complaints on the other hand, could be a bit subjective, since it represents things the customer did not like, and took the time to tell you about. Even one complaint could be a tip of an iceberg, since not everyone takes the time to tell the company about something they did not like or something that did not work with respect to the product or service.

Understanding customer behavior patterns

There are things which customers tell an organization without actually saying anything. This may sound paradoxical, but looking at service usage patterns, product buying trends, calls to question product features, or even website traffic nuances will provide a deeper understanding on customer likes, and how they use the product or service. This data mining approach can yield great insights, and should not be overlooked, since in many cases, patterns observed though this method may be more powerful than any of the other methods discussed.

Having a customer focus will ensure that an organization is always thinking ahead in terms of improving its processes, and innovating its products and services. Losing sight of the customer can be a fatal mistake for any business. Multiple feedback channels from customers will ensure that correctly prioritized issues are being worked on for the future.

Quality Metrics

The word used for measurement is called 'metrics'. Metrics represent numbers which measure the conformance of a product or service to a goal. They provide a view of how a department, organization, or process is operating and can be used to determine if a reaction is needed to any deteriorating circumstance. Quality metrics encompass all operational and customer feedback information. In that sense, they are a pulse of how things are running. When used in conjunction with financial, marketing, and HR metrics, they form the basis for a balanced scorecard and display the landscape of a firm's operations.

Having a customer focus is good, but may not be enough to ensure that business performance will improve in a concerted manner. Appropriate measurements must be in place to determine if work is yielding good results. All groups, departments, organizations, and individuals must work towards these metrics to gage success. Although all measurements should be customer centric, specific groups can have their own metrics, related to their job functions.

Use of customer focused metrics

The selection of what to measure is of paramount importance as you are only as good as your measurements. As the old saying goes, 'Tell me how you'll measure me, and I'll tell you how I'll behave'. If something is not being measured, no one will know if performance in that area is improving or deteriorating. One of the main challenges in selecting metrics is to ensure that they are from the customers' point of view. This 'outside inward' approach helps isolate the important parts of the process. In general, there are two basic types of quality metrics: Operational and Qualitative.

Operational metrics

These metrics show how a process is performing over time, and if it running without any 'special causes'. Special causes are external imbalances or defects which create excessive failures. Operational metrics should reflect these imbalances quickly when they appear and actions should be taken to fix them immediately. As discussed in the previous section, operational metrics should be as customer focused as possible. If the process is showing a high success rate, but customers are unhappy about the product or service, then the metrics should be made more meaningful. This is a great way to link operational performance to customer satisfaction.

In an ideal setup, if all the operational metrics were to perform as desired, then the customer should have a great product or service experience.

Qualitative metrics

These are a reflection of customer feedback in qualitative terms. They include survey results, customer verbatims, or unsolicited customer feedback.

To understand how a process works, a combination of operational and qualitative metrics should be employed. Using just one type of metric may not be enough to pinpoint an issue. For example, a process may be running well with no special causes, but customers may still be dissatisfied, due to something in the process that is not being measured. Customers may also provide feedback about things they expect, which this process may be unable to deliver. This could loop back into the possibility of obsolescence and should be an indication for the need to innovate.

To summarize, quality metrics form the backbone of any organization's operations and should be used as a tool to improve processes which in turn should increase customer satisfaction.

Commonly used metrics

Each organization will have its own industry and company specific metrics. There are some metrics which are commonly used across various industries, and every effort should be made to use them, when applicable. Some of these metrics are:

Defects per unit or DPU: This is the average number of defects in the total number of units inspected. A 'unit' is the item being inspected, such as a door handle, or a service being delivered. DPU is calculated by dividing the total number of observed defects by the total number of units in the sample.

eg: 20/100 = 20%

DPU is also used when there may not be a separate parts but a continuous flow of a product, such as textile, metal sheets, etc.

Parts per million or PPM: This is the total number of defective parts divided by total parts. The result is then multiplied by a million.

eg : 20 / 5000 = .004 x 1,000,000 = 4000 PPM

PPM counts each part as either defective or not defective. A part that is not defective has no defects at all while one that is, may have one or more defects. In the door handle example, the number of errors that occur on a door handle does not matter. There could be one defect, or 10. In both cases, that handle is considered as one defective unit. PPM is used when processes reach levels where a percent does not provide enough discrimination, or when very high levels of quality are desired.

Chapter 5: Get the Details

The next morning was one of paradoxes. On one hand Jeff knew that his follow up presentation on Larry's project was not going to go well. He didn't understand where part of the data was coming from, how the numbers were calculated, and what the source authenticity was. On the other hand, he was feeling good about what he had learnt so far and how he was going to manage his work going forward. The list of things that he had to accomplish was getting longer, but these were things he wanted to work on. They would get him to the desired state.

Jeff was prepared for his team meeting. He was ready to share what he had learnt. "From the management point of view, it just makes sense," Jeff told himself. "Encouraging narrow thinking will not get us where we want to be as an organization. Strategic planning with effective communications will." In the ten minutes he had before the meeting, Jeff began to jot down additional thoughts he was going to present to the team:

Cross train team members: Cross training has several benefits. It not only provides back-up, but also allows you to identify each person's strengths, and then allocate work based on what they do best. From an individual's point of view, cross training gives them a chance to learn new job skills, and eventually enhance their careers. Coach and train all to ensure that each one understands how to perform their job effectively. This can be done in many ways, such as:

a. Have one team member work with another to impart their skills. This is also called 'shadowing'. The trainer should be well versed in the activity being taught.
b. Train the person yourself
c. Have the person be part of a cross-functional team with minimal responsibilities with the intent to imbibe and report-out on the team's progress.

Use a combination approach in which you may do part of the training, and have another team member help with other parts. To ensure that each person is well trained, list out skills and knowledge required to do the job well. Then identify areas where improvements may be needed.

Jeff presented his thoughts to the team. He started with the basic expectations he had for all of them, and then talked in-depth about cross-training. Knowing that not everybody would like to train another person with the skills needed to do their own job, he addressed the case with care. "It is something we have to do to protect ourselves," he said. "We are all human, and could fall sick, get into an accident, take a leave of absence, or as we do so often, go on a vacation. A sound backup system will allow work to continue as normal. This can be done with an each one, teach one system."

Jeff was soon answering questions such as "where are we going to get the time to cross-train another person? Does this mean that one of our jobs is at stake?" For each question, the answer had the same theme; cross training is good for you, good for me, and good for the company. It is a win-win situation. Unfortunately, since we are in a fire-fighting mode most of the time, cross-training gets pushed aside. A little investment of time up-front goes a long way in running the business right. Besides, more knowledge empowers people to determine where their interests lie.

Jeff felt good after emphatically telling his team that in order to progress, they had to learn, and more importantly, they had to teach. To his surprise, all this talk about expectations and cross-training seemed to settle well with the team. Going into the meeting, his hope was their acceptance. In the meeting, he realized that they were bringing up valid points which he had not thought of himself. "Should we cross-train with someone outside our group, to gain a skill set that we do not possess, but would be great to have?" "Should we develop a training library, where reference material related to the job on hand is readily available?" These kinds of questions made Jeff feel a lot better. After the meeting, he quickly went over and told Dean about the discussion he had just had with his team. Once again, true to his style, Dean said "your thoughts are indeed in the right direction. What you now need are the details. A good cross-training plan should be specific, detailing the tasks, people, and level of competence. A matrix would describe this well. As you move along to reinvent your job, this is a task you can undertake and implement. Remember, the matrix should be a live document, and should be updated regularly. More importantly, nothing you do will be effective until you look at the human side of the equation. Teamwork should drive cross-training."

Desired State

You have created an effective backup system to cover for any team member's absence. This backup plan is documented. You can operate with a reduced workforce during the time you do not have a replacement and will still accomplish tasks your team has been assigned.

Jeff knew that cross-training would work only if there was a sense of teamwork within the group. If each one wanted the other to fail, why would they want to cross-train? Jeff started reading material related to teamwork, and how it should be used to improve quality.

Teamwork

Teamwork plays an integral role in improving the collective output of an organization. Better teamwork results in improved quality of products or services. In the past, Japanese companies instituted 'quality circles'; groups of employees who would get together and talk about quality related issues to improve processes. Quality circles have had their metamorphosis into several versions of team building exercises.

Teams are usually cross-functional in nature and their goal is to work together to accomplish an objective. A factor that plays a big role in determining the success or failure of a team is the organization structure that people report into. A well designed organization structure encourages teamwork, and breeds success.

High performance teams

Creating high performance teams will promote better teamwork, and improve results. Each team must have a clear leader and each member must have a good understanding of their roles. Teams can consist of departments or groups or they could be cross-functional in nature. The focus must be on the team's performance, and not that of any one individual. The sum of the individuals must be less than the output of the team for it to be successful. Individuals must have good accountability for their responsibilities, and the leader must have the ability to chart the course of their work. To determine success, it must be decided early on how to measure performance. These measures could be qualitative or quantitative in nature. Each team should have its own life-cycle. This means that it should start with an objective and end when that mission is accomplished.

Team Stages

Bruce Tuckman proposed four stages for teams, which are Forming, Storming, Norming and Performing. He suggested that these stages are necessary for teams to be successful. In the forming stage, teams get together, and form their own identity. During storming, viewpoints from individuals are brought to enrich the team's abilities. Norming allows people to settle down and adjust their behaviors to be consistent with overall goals. Finally in the performing stage, the team begins to deliver results.

Chapter 6: Down in the Dumps

Jeff was a great believer of delegation, and that was what had got him into trouble. He had given Larry a free reign which caused him to be out of touch with how work was done. He began to realize that there was a difference between delegation and lack of management. Jeff had already confided about his lack of knowledge to Dean. "Dean is smart," Jeff thought. "It made sense for me to tell him the truth before the meeting. If something goes wrong, he will be on my side, rather than be against me along with the rest of the audience."

Jeff remembered how Dean had been astonished when he had told him how Larry got things done, without his knowledge of details regarding the how, when, where, what, and why of it. "Let's get this straight," Dean had said in an authoritative voice. "There is delegation, and then there is delegation. Asking an employee to get a task done does not mean that you have no clue about it. Being a manager puts the onus on you to understand what you want done, by when, and in a broad sense, how. You don't have to micromanage activities, but you should be a guide. Show them the direction, and let them find the path." Jeff remembered those words, which were both powerful and insightful.

Dean had asked Jeff to meet him before the presentation to go through the slides. Instead of reviewing the content, Dean continued from where he had left off. "The point is," said

Dean "you have to empower your team to take on tasks and complete projects. This will help you work on bigger and better things. In order to empower them, you have to first understand the system well, and remove high level roadblocks that they may encounter. If you do not engage in this roadblock blockbusting upfront, your team may not be able to counter pitfalls downstream and will not succeed. As a result, they may end up finding other ways to make things work for them. In some instances these may be good workarounds. In other cases, they may be time consuming, and have diminished returns. If you understand what they do, then you can remove system inefficiencies without getting into the nitty-gritty of things."

Dean then talked about some specifics. "Work should be reviewed periodically for status updates. Initiatives that team members work on should be reviewed for alignment with overall goals. The best way to see the results of a project is depict it with a quality tool. The tool should be selected based on the type of data being reviewed. Quality tools are a quick and easy way to determine what the data is saying."

As Dean finished, Jeff had begun to tune his mind out and was thinking of the next step. Yes, his boss had seized one more opportunity to lecture him, but he had learnt another valuable lesson. He knew he did not have many more chances. He knew that the clock was ticking in Dean's mind on his abilities, and he had to quickly show what he was capable of. It was now or never. Jeff did not remember what he said while leaving, but single mindedly got back and translated the conversation he had just had into plain

English. In order to learn, he was paying with his reputation. He didn't have much of it left, but he didn't have a choice either.

Empower: The team should feel empowered to make decisions. This philosophy of empowerment should be coupled with a system in which roadblocks are removed, work done fits into the bigger scheme of things, and status updates and results are available using established quality tools. As the goalkeeper, each team lead needs to look at the landscape of projects on hand, understand them, and drive for results.

Jeff's vision was a bit blurry, and his head was spinning. He knew that he had to improve, but repairing the damage with Dean was going to be difficult. "I better start looking at the classifieds," he told his wife that evening. "I might need it sooner than I think."

Desired State
Each person on the team feels empowered to make decisions which fall within their range of work to accomplish enterprise goals. These decisions are based on established criteria. The status and results of these projects are communicated with quality tools.

Jeff started looking into the 'Quality Tools' that Dean had talked about. He realized that some of these tools were simple, yet powerful and were referred to as 'Basic Quality Tools'. These tools show the state of a process or the results of a study. They remove the subjectivity out of decision making. They are represented as status charts, and the decision about which chart to use depends on the scenario on hand. In some instances, it could also be a matter of choice. It is important to know how to build and interpret some of the common charts, as they provide a view of what's going on, and help make decisions. Going with the theme of 'trust but verify', Jeff wanted to learn more about these quality tools, so his delegation would be more systematic, and not just 'passing over' work, without knowing anything else about it. He wanted his team to use these tools to report out their work making it all data driven. He reviewed some commonly used quality tools to gain that understanding.

Quality Tools

There are several quality tools which present data in an easy to understand way. Some of these are data analysis tools, while others are problem solving tools. Data analysis tools show the status of existing processes and are used to prioritize work. Problem solving tools are used to determine root cause of issues. These tools show data in a compact and easy-to-understand way by using averages, percentages, etc. They give more information about a process than individual data elements do. These tools can also be used to review results of projects. The following are some of the commonly used quality tools:

Data analysis quality tools:
1. Check sheet or tally sheet
2. Bar chart
3. Pareto chart
4. Histogram
5. Dot plot
6. Stem and leaf plot
7. Box and whisker plot

Problem solving quality tools:
1. Cause and effect diagram
2. Flow chart
3. Problem definition tree
4. Scatter plot

Check sheet or tally sheet

Check sheets or tally sheets are tools for recording data in an organized way. They are used for understanding sources of data, and group it in a meaningfully. They help determine which problems should be prioritized for action.

Check sheets can be customized to collect and analyze data based on the information needed. Some of the common types of check sheets are:
a. Plain frequency check sheets
b. Nonconformity by location check sheets
c. Nonconformity by causes check sheets

Check sheets help uncover how numbers are distributed from a given data source. They show centering, and dispersion of data, but do not show the order in which data was collected. The lack of time sequencing of data collection does not allow this tool to determine process stability. So, check sheets are geared towards data collection, and used as an input for problem solving.

The example on the next page shows a check sheet in which data is collected for scratches, dings, and dents on a vehicle door. The location of each defect is marked on the diagram, and its tally is counted at the top. At the end of the shift, the supervisor will be able to see the pattern of these defects. There are strong clues on this check sheet which point towards a handling issue causing the scratches and probably a machine interference issue causing the dings. These will have to be investigated and validated. This check sheet provides a great first step in identifying problems to root cause.

Sample Door Metal Defect Check sheet

Workstation Number:

Date/Shift:

Batch Number:

Inspector Name:

Number of Doors Checked:

Defect Type	Symbol	Count
Scratch	○	
Ding	△	
Dent	▭	

Bar chart or column chart

Bar charts are made up of bars plotted horizontally, and column charts are columns plotted vertically on a graph. In a bar chart, each bar represents the category and should be labeled accordingly. The height of the bar indicates the size of the category. Besides the layout, there is no other difference between a bar chart and a column chart.

These charts are commonly used to understand impacts of categories. In the charts on the next page, we see that doors have the largest number of defects, while fascias have the least. The two charts look quite different, but they are actually representing the same data. The different scales tend to make the charts appear different.

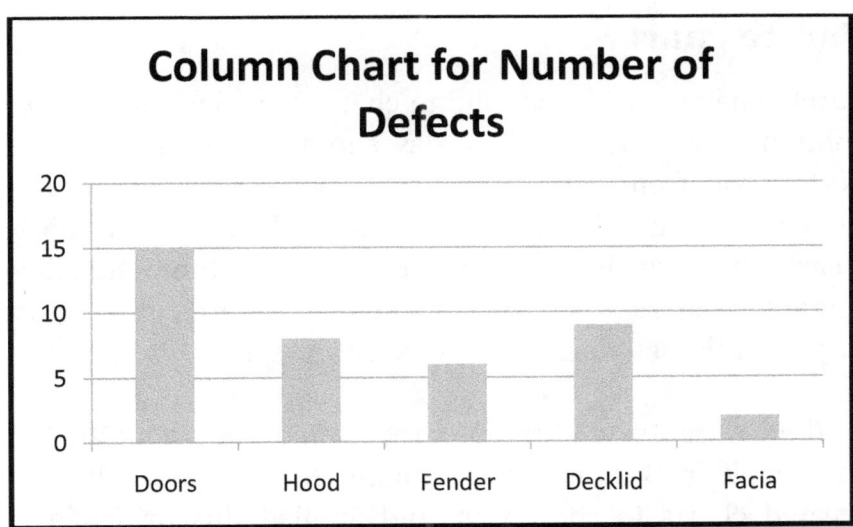

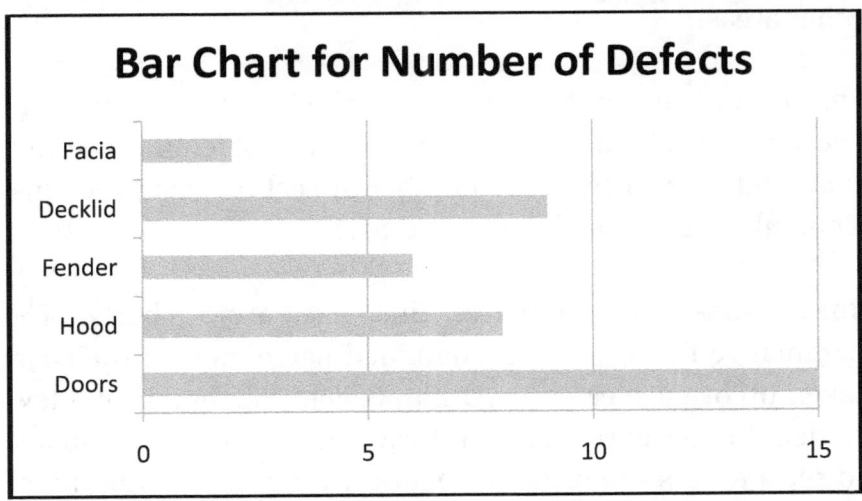

Pareto chart

Pareto charts are bar or column charts in which the bars (or columns) are arranged in a descending order from top to bottom or from left to right. They highlight the most important issues for a process or product. Pareto charts clearly show which issues are the most important, and allow limited resources to be channeled towards the most important issues which need to be addressed.

In the 19th century, Vilfredo Pareto found that about 20% of the people in Britain owned roughly 80% of the wealth. He created charts to show this, and applied this principle to other areas.

Dr. Joseph Juran brought this principle to quality, and showed that about one fifth of quality problems accounted for about four-fifths of the number or cost of such problems. This rule is also called the '80-20 rule'.

Juran added a cumulative line on Pareto charts. The cumulative line tracks the combined percentage of the chart taken up by each category. Juran coined the terms 'vital few' to denote the more significant categories, and 'trivial many' to refer to less important categories. Now, the 'trivial many' is replaced by the term 'useful many', since one can never be sure if anything is really trivial. However, to maximize return on fixing problems, we need to concentrate on the vital few problems first.

Pareto charts can be made for different levels, to hone in on details than higher level Pareto charts. They can also represent different perspectives. For example, one chart could track problems by root cause, while anther could identify reasons for customer dissatisfaction. Another important feature of Pareto charts is that they can provide the 'before and after' pictures of processes. This helps evaluate success in problem solving efforts.

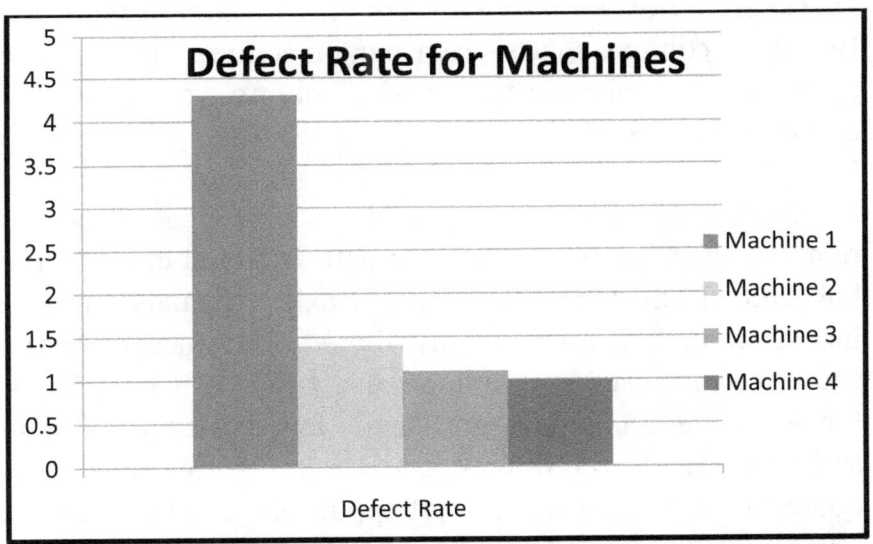

In the Pareto chart above, we see that machine one has the highest defect rate of over 4 percent. The other machines have defect rates less than half of machine one. This chart tells us that machine one should be worked on the reduce defect rates, since it is the highest contributor.

Histogram

A histogram is similar to a bar or column chart, with a few differences. Usually there isn't any space between the columns in a histogram. Each column shows how many times something occurs. The column label can have a single value or a range of values, and the height of the column indicates the size of the group defined by the column label.

Difference between bar charts and histograms:
In bar or column charts, each bar or column represents a category. In histograms, each column represents a quantitative variable.

A histogram's sknewness should be observed. Skewness refers to the tendency of the data pattern to fall more on the low end or the high end of the x-axis. In many cases, a histogram may be normally distributed. This means that the shape of the histogram is that of a bell. With bar charts, however, the x-axis does not have a low end or a high end since the labels on the x- axis are categories, and not quantitative. So, skewness is not applicable for a bar chart.
Histograms show variation of data. They have ranges, each indicating a zone within which measurements occur. The height of each range is determined by the count of individual measurements within that cell. Thus, histograms show centering and dispersion of the data. However, since they do not show value over time, they cannot provide any insight into process stability.

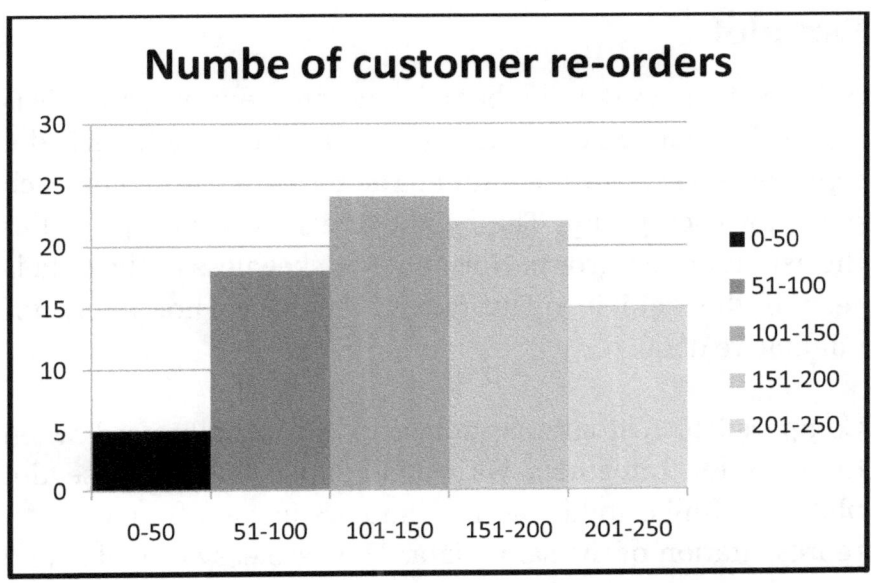

In the above histogram, ranges on the x-axis are in units of 50. The y-axis shows the number of times observations fell into each of the ranges. The pattern of the histogram shows that it is roughly normally distributed. Most customers had 101 to 150 reorders. Only 5 customers had re-orders in the zero to 50 range. These customers could be contacted to determine why they did not re-order enough. Similarly, the 15 customers who re-ordered in the 201 to 250 range could also be contacted to find out what they like about the product.

Dot plot

A dot plot shows the number of times an event occurs. This is shown by plotting dots on a graph. In a dot plot, each dot represents a single observation. The dots are stacked on each other for each group. The height of each column quantifies the extent of that group. Normality or skewness of the data is seen if the data is quantitative. This does not apply for qualitative data.

Compared to histograms, dot plots are usually used when there are few categories, with small data sets. Otherwise, dot plots are similar to histograms, and are just a different visual representation of the same data. They are easy to make, and are sometimes created manually too.

Dot plot Example

Here is how a dot plot looks like, and how it should be interpreted. If a group of people are asked to select their favorite mode of transportation from a list of choices, then their choices can be summarized in a dotplot, as shown on the following page.

```
*
*       *
*       *
*       *               *
*       *       *       *
*       *       *       *
*       *           *   *   *
*       *   *   *   *   *   *
*       *   *   *   *   *   *
```

Car   Air   Rail   Bike   Walk   Bus   Boat

Each dot represents one person, and the number of dots in a column represents the number of people who selected the mode of transportation associated with that column. For example, car was the most popular mode, selected by 9 people, followed by air. Boat and rail were the least popular, selected by two each.

The categories in this example are qualitative, so normality or skewness should not be interpreted from this data. If quantitative data exists, like in the example below, one could observe normality or skewness.

The dot plot below shows the number of televisions owned by each family in an apartment complex.

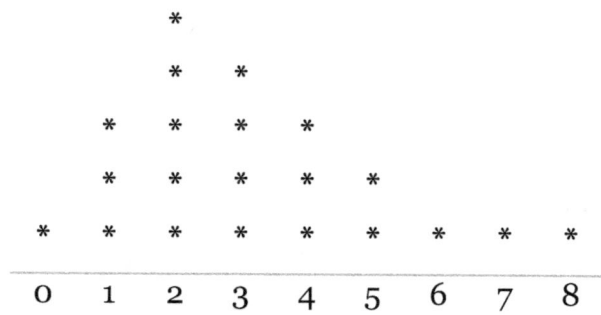

Here, the distribution is skewed to the right, with no outliers. Since most of the observations are on the left side of the distribution, the distribution is skewed to the right. None of the observations appear far to the right, so there are no outliers.

How to Compare Distributions

When comparing data sets, focus on three features:

Center: The center of a distribution is the point where about half the observations are on either side.

Spread: The spread of a distribution shows the variability of data. If observations cover a wide range, the spread is large. If observations are in a narrower band, the spread is small.

Shape: The shape of a distribution is described by symmetry, skewness, number of peaks, etc.

Stem and leaf plot

A stem and leaf plot is used to display data, usually from a small data set of 50 or fewer observations. The stem plot below shows test scores for 30 fifth graders.

Stems	Leaves
100	
90	1 3 5 6 9
80	2 4 4 4 7 8 8 9 9
70	1 2 2 3 3 5 7 7 8
60	7 9
50	1 4
40	0 9
30	1

Numbers on the left are called the stem and those on the right are called leaves. In the above example, the stem is in sets of tens.

Some stem plots include a key to help the user understand the data better. For example, in the example above, one should be able to describe the distribution of test scores. Most of the scores are clustered between 70 and 90, with the center falling in the neighborhood of 80. Scores range from a low of 31 to a high of 99. For this stem plot, the range is 68, and the median is 77.5

The range is equal to the largest number minus the smallest number. The largest number is 99, and the smallest is 31, so the range is 99-31= 68.

The median is equal to the middle value in the data set. Since we have an even number of values 77 and 78, their average is (77 + 78)/2 or 77.5, so the median is equal to 77.5

Box and whisker plot

A box and whisker plot is also called a box plot. Data is split into four parts, called quartiles. A central box forms the body of the plot, and contains data from the first (Q1) to the third (Q3) quartile.

Inside the box, a vertical line is drawn at Q2, which is the median of the data. Two horizontal lines, called whiskers extend from the front and back of the box. The front whisker goes from Q1 to the smallest non-outlier in the data set, and the back whisker goes from Q3 to the largest non-outlier.

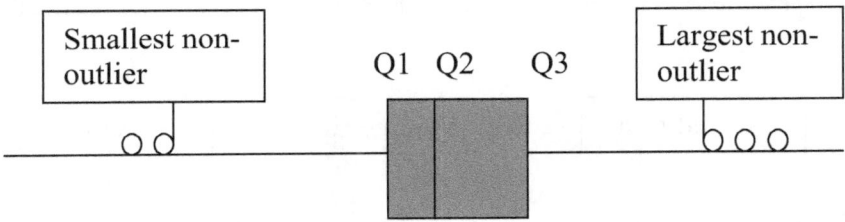

How to Interpret a Box plot

The median is the vertical line that runs through the box. Box plots also display two common measures of variability or spread in a data set.

Range: A box plot shows the distance between the smallest and largest value, including any outliers. In the box plot above, data values range from about -700 (the smallest outlier) to 1700 (the largest outlier), so the range is 2400. If you ignore outliers, the range is the distance between the opposite ends of the whiskers which is about 1000 in the box plot above.

Interquartile range (IQR): The width of the box shows the interquartile range (Q3 minus Q1). In the chart above, the interquartile range is equal to 600 minus 300 or about 300

Finally, box plots often provide information about the shape of a data set. The examples below show some common patterns.

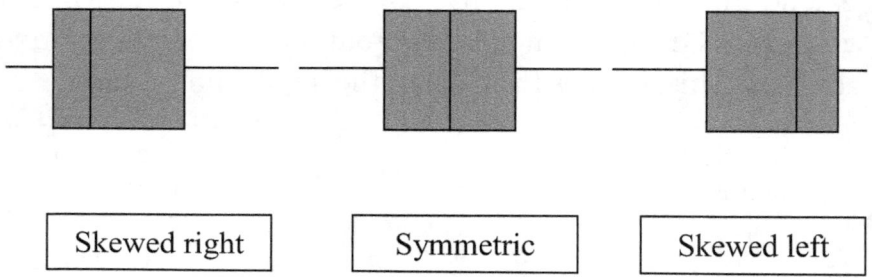

| Skewed right | Symmetric | Skewed left |

Each of the above box plots shows a different skewness pattern. If most of the observations are concentrated on the low end of the scale, the distribution is skewed right; and vice versa. If a distribution is symmetric, the observations will be evenly split at the median, as shown above in the middle figure. Consider the box plot below.

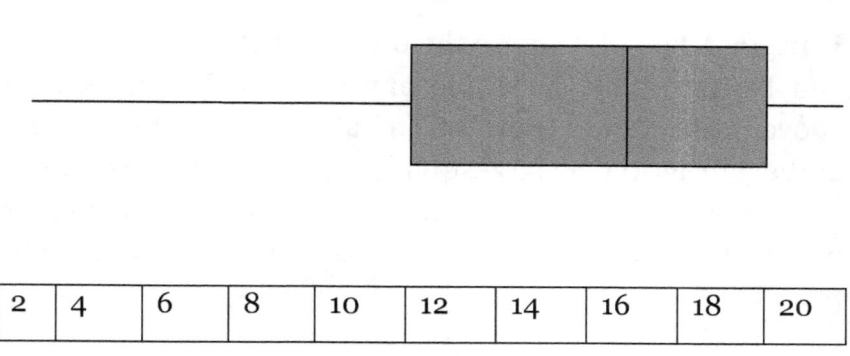

| 2 | 4 | 6 | 8 | 10 | 12 | 14 | 16 | 18 | 20 |

For this box plot, the interquartile range is about 6. Most of the observations are on the high end of the scale, so the distribution is skewed left. The interquartile range is the length of the box, which is 18 minus 12, or 6. The median is the vertical line running through the middle of the box, which is roughly centered over 16. So the median is 16

Cause and Effect Diagram

A cause and effect diagram is also known as fishbone diagram, or Ishikawa diagram. It is similar to a tree diagram, and is used as a group brainstorming tool. Its use is to record information on potential contributors to a given result. The fishbone diagram helps to sort causes into major areas. This allows for structured brainstorming. Like the flowchart, the cause and effect diagram may be used in problem solving or planning activities. Participants contribute their ideas of likely potential causes which are recorded on lines attached to the main cause. Only those knowledgeable about the process should be involved in creating these diagrams.

Fishbone Diagram for Scrap Creation

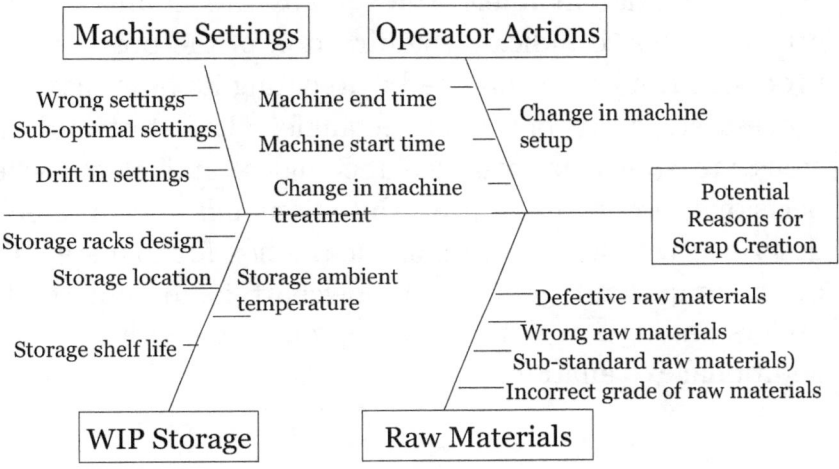

In this fishbone diagram, we see that the potential reasons for scrap creation are listed on the 'bones' of the fish, or 'branches' of the tree. Each bone or branch has a sub-branch which provides additional categorizations.

Flowchart

Flow charts show the various steps of a process, in a step-by-step sequential manner. These steps are connected with arrowed lines to indicate the direction of the process. Most processes can be documented by creating its flow chart. The process steps are boxed in rectangles. The start and stop steps are shown by rounded rectangles, and places where decisions occur are shown by diamonds. Each step is described with a few words inside the box for that step. One of the important goals of a flowchart is to impart clear understanding of the basic processes without much additional explanation.

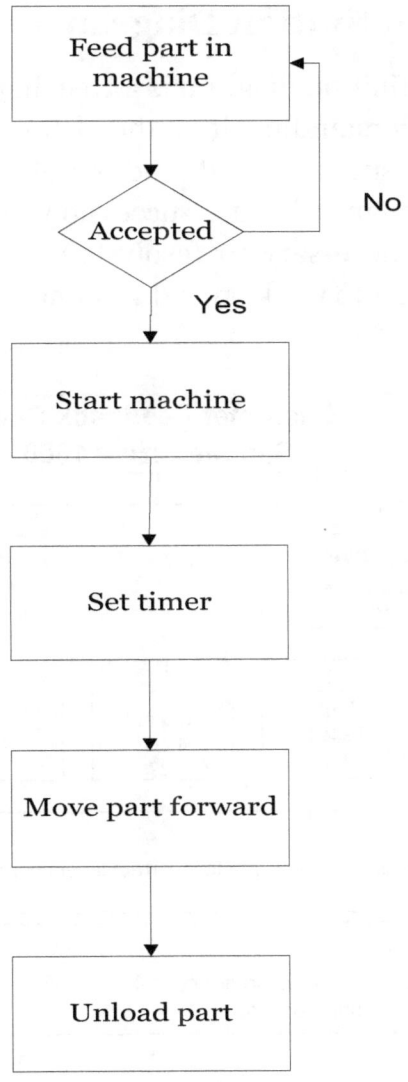

In the above flowchart, the processing of a part in a machine is explained in a step-by-step manner. There is a diamond shaped decision box, which shows that if the part is not accepted by the machine, then the process begins from the start again. The last step is to unload the part.

Problem Definition Diagram

A problem definition diagram is a tree diagram that provides a deeper understanding into the details of a problem. It refines the issue on hand and breaks it down into its elements through a logical succession of steps. This helps drive actions necessary to resolve the issue. It does so by identifying what to work on and what not to work on.

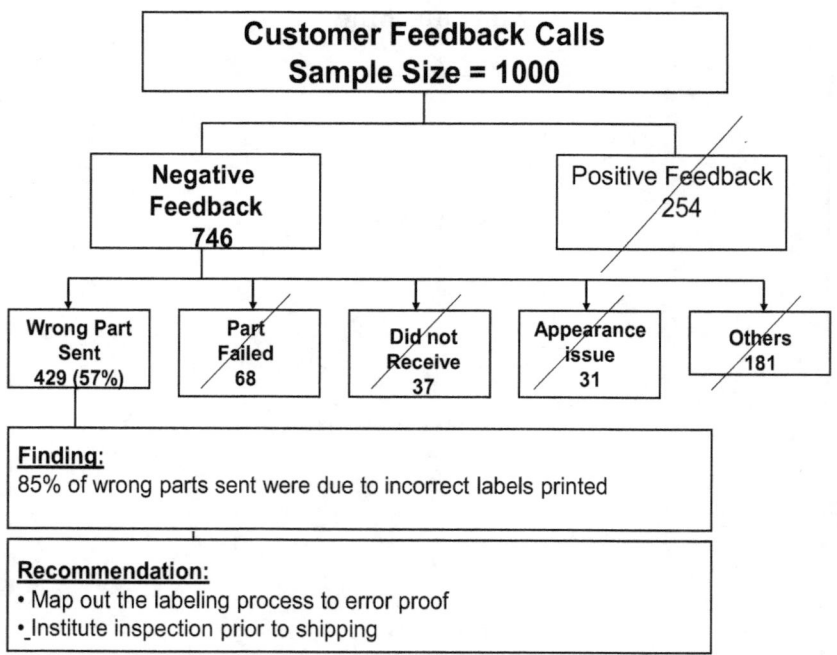

In this problem definition diagram, we see that customer feedback calls are categorized based on their nature or disposition. They are further broken into what customers are calling about. The area to focus on is to determine root cause for the reasons why wrong parts are being sent.

Another tool similar to the problem definition diagram is the 'why-why' analysis. This also uses a tree diagram to get to the root cause of a problem. Why-why's can be done on items chosen from the cause and effect diagram. It focuses on a particular cause for a problem and asks why that cause exists. There could be more than one answer to each 'why'. Each answer should be evaluated and their reason investigated. The number of levels of whys needed to get to the root cause will vary depending on the complexity of the problem. It has been observed that five levels are usually adequate.

Scatter plot

A scatter plot consists of dots plotted on a graph. Each dot represents the co-ordinates of the x and y axis. It helps determine if a relationship exists between the variables of the x and y-axis.

Scatter plots show if the change in value for one characteristic influences a change in value of the other characteristic. If a scatter plot shows a relationship between two variables, it does not definitively mean that one causes the other to change. The two could be impacted by a third variable. The table below shows the height and the weight of six players on a basketball team. The same data is then displayed on a scatter plot.

Height	Weight
69	179
71	185
73	190
75	215
77	220
79	223

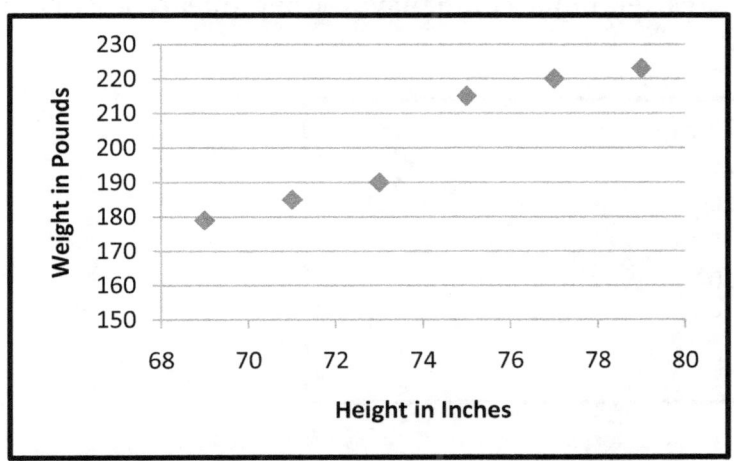

Each player in the table is represented by a dot on the scatter plot. The first dot represents the shortest, lightest player. From the scale on the x-axis, we see that the shortest player is 69 inches tall; and from the scale on the y-axis, we see that he/she weighs 179 pounds. Similarly, we can read the height and weight of all players on the scatter plot.

Data patterns in scatter plots
Scatter plots are used to analyze patterns in data. Some of the things to observe in a scatter plot, which will provide clues on the relationship between two variables are reviewed on the following page:

Pattern: Scatter plots could have either linear or curved patterns.

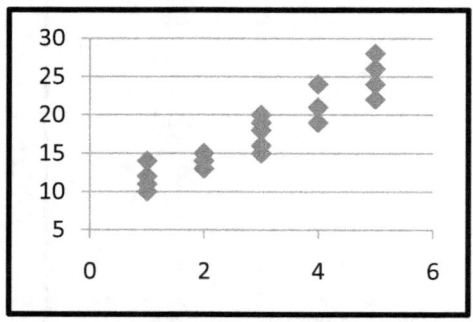

Linear

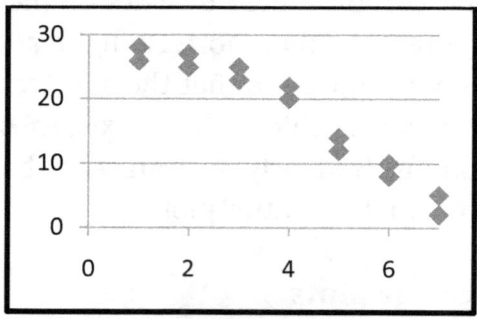

Curved

Slope: Is the slope of the imaginary line passing through the dots positive, negative, or flat? A couple of examples are shown on the following page.

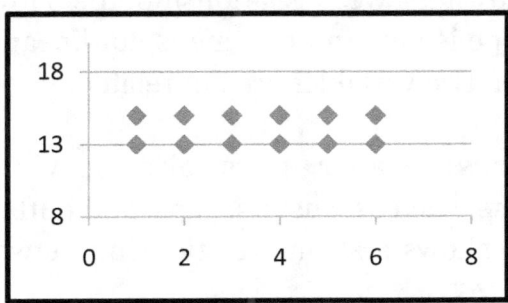

Flat slope

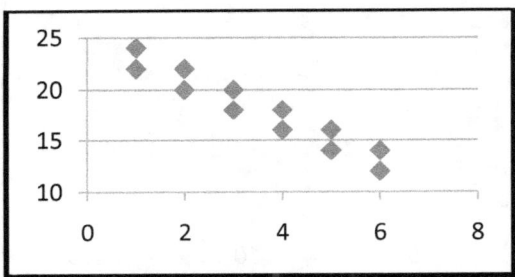

Negative slope

Degree of scatter: If dots are widely spread, the relationship between variables is weak. If the dots are concentrated around a line, the relationship is strong.

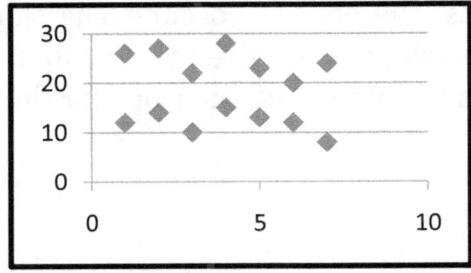

In the example showing a weak relationship, the pattern also shows that the slope is flat, and the line is not linear. This is usually found when two variables are not related.

Unusual features: Scatter plots also show unusual features in data sets, such as clusters, gaps, and outliers. The scatter plot below shows a strong relationship between two variables.

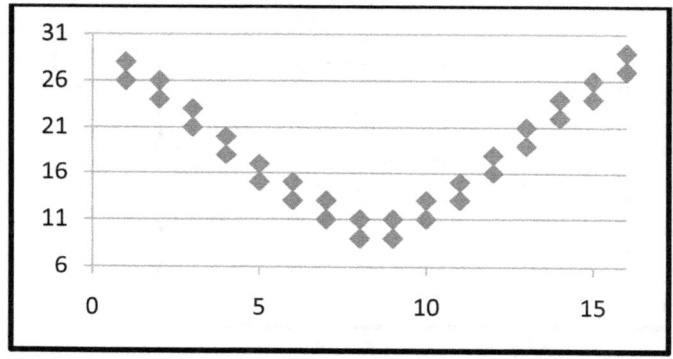

The relation is strong because dots are tightly clustered around a line. Note that a line does not have to be straight for a relationship to be strong. In this case, the line is U-shaped. Across the entire scatter plot, the slope is zero. The slope in the first half of the scatter plot is negative but in the second half, just the opposite occurs. The slope is positive there. When the slope is positive in one half of a scatter plot and negative in the other half, the slope for the entire scatter plot is zero.

Chapter 7: Optimism

In spite of the negativities of the previous day, Jeff was upbeat the next morning. "You may have not realized it, but you have gained a lot in the past few days," his wife had told him. "Besides, in order to learn what you did, going through this process is a small price to pay. The knowledge you have acquired is priceless." Armed with his new found optimism, Jeff decided to continue with the pursuit of sound management practices complemented by an equally sound data driven approach. He knew that things will begin to come together and what he had written down and learnt would soon be a 'mantra' for him, and for others. One of the things Jeff wanted to ensure was to have the ability to access current actions and project summaries. Although he assigned projects to his team, he soon lost track of them. This had caused him a good deal of embarrassment with Dean. If he followed through with everything he had written down in the past few days, the icing would be to have an up-to-date summary of what was happening on each project. At 6:00 a.m., over a cup of coffee, Jeff added the following to his notes:

Summarize: At any given time, you need to know a snapshot of various tasks being performed. It is not good to be asked how a task is progressing and not know the answer. The main reason leaders do not know a summary of what their teams are doing, is because work load gets overwhelming to manage. The solution to this is to develop a grid that clearly lists each project, and shows how it is progressing. Each project should be summarized with words and numbers so that there is adequate quantification. Every team member should be able to update this information on a daily/weekly basis to provide visibility on the progress.

"If I were to show Dean what I just wrote down, what would he say?" pondered Jeff, now trying to preempt Dean's comments. "He usually focuses on the details. In this case, it would mean, how would the summary look? What metrics would you focus on? How would data be represented? Would there be an action items list, with responsibilities, target dates, et al?"

Jeff soon began to write up an action items list, target dates and important metrics for each of the projects his team was working on. He started to summarize many of his projects. He did not have much time, since he knew that Dean would be walking in soon, but he at least wanted a rough draft, so that he could use that as a basis for discussion. Since he did not know some of the details of the projects his team was working on, he left those places blank.

With some fodder for talk, Jeff was ready to meet Dean that morning.

Dean looked impressed initially, but as soon as he realized this was only a proposal, he lost some of his interest. "I like what you have done Jeff, but it would be great to have the real thing; something that I could look at, and use as a basis for a decision." Jeff knew that Dean would always want something more than what he had to offer, and took the comments in his stride. Putting himself in Dean's shoes now, he felt that proposals were good, but could sometimes be time buyers. It would have been great if he had taken the time to show Dean some real data that would help him understand how things were progressing on various fronts.

Desired State
Data is collected and statistically summarized. Action items with target dates are available and updated by team members regularly.

The field of data collection, analysis, and summarization is a large one. If data collected is tainted due to any reason, then the decision made from that analysis would be suspect. So, this process should be based on a certain methodology to set the basis for data analysis. An important part of data analysis is the understanding of probability distributions. If the underlying distribution of a data set is understood, then its analysis can be done accordingly.   These probability distributions should be kept in mind while collecting data too. They form the basis for several statistical calculations, and understanding them would help the team make better decisions.

Once data is collected and analyzed, summarizing it in a meaningful way to form the basis for decision making is critical. Use of basic statistics helps summarize data and talk about it intelligently. Jeff wanted to learn more about data collection, analysis, and summarization. He knew that once summarization was complete, it could be represented with some quality tools he had already learned. He began to get his feet wet with the three areas of data collection, analysis, and summarization.

Data Collection

Data collection is a very important phase of a project. It forms the basis for good results. The analysis is only as good as the data itself. So, the thought process used to determine what data should be collected is very important.

Data Stratification

This is an approach which can be used for data collection. This model will help in collecting useful data, rather than any data that is readily available. When data is stratified, it is broken down into layers, or 'strata'. The questions who, what, when, and where represent these layers. By answering each question, it becomes possible to move closer to determine details of the issue on hand.

Let us consider a problem, in which the product return rate at a company has spiked since the past month. In this scenario, if the question 'who' is asked, it will help determine which supplier's parts are being returned at a higher rate. It could also point to the type of customer returning the products. 'What' will help point out if the problem is related to a particular model. 'When' will lead to determine if the problem had started since a new product or service launch. 'Where' will direct attention to see if returns are coming in from a certain geographical region.

Example of data collection method

A computer company's laptop computer line is experiencing a high percent of returns over the past year, and the rate continues to move higher. The investigation team needs to collect data to define the problem.

The team lead decides to use the data stratification approach to collect data. The goal is to determine the cause of the problem that is affecting the computers.

The first question the team asks is 'who?' The team wants to know which type of customer is returning computers the most. If a particular type of customer can be identified, it will lead them closer to the root cause of the problem. By asking 'who' the team finds that majority of the users returning laptops are high end users, utilizing gaming systems, and multimedia.

The second question the team asks is 'what?' That is, what are the main reasons computers are being returned? An investigation reveals that reasons range from slow speed to long boot up time. All reasons were performance based.

The third question the team asks is 'when?' During what time period are majority of computers returned? In asking when, the team searches for patterns that might reveal a particular manufacturing flaw or some other kind of defect, that cause computers to begin functioning at sub-standard levels. Team members find that the rate of return increased about a year ago. That's about six months after the new model was introduced.

The fourth question the team examines is 'where?' Where are computers experiencing a high rate of return? Is it a particular retail chain, city, state, or region? Also, where on the computer is the problem occurring?

This kind of questioning and data collection will lead the team closer to understanding why returns have increased. It is the first step in moving closer to the root cause.

Basic Statistics

Basic statistics covers areas that provide information on the measures of central tendency, and measures of variation. The most commonly used measures of central tendency are mean, median, and mode. The commonly used measures of variation are range, standard deviation, and variance. These six measures provide enough information to understand the central tendency and variation of a given data set. Also, basic statistics introduces the idea of data distribution. This is represented in the form of some common distributions such as normal, skewed, and bi-modal.

Measures of central tendency

As stated above, the three commonly used measures of central tendency are mean, median, and mode

If the mean, median, and mode are equal, then the distribution is symmetrical. In the real world, these distributions are not perfectly symmetrical, but may approach symmetry.

Mean

The mean is the most commonly used measure of central tendency. It is also called the average. The mean is the sum of all observations divided by the number of observations.

Example:

Find the mean of the following data set:
1,2,2,3,3,3,4,4,4,4,5,5,5,6,7,8,9,10

Here the sum of all values is 83, and there are 18 numbers. So, 83/18 is 4.61, which is the mean.

Mean=total of all values/number of values

In many cases, the mean may not be the best measure of central tendency. Extreme values can pull the mean in one direction from the center. This distorts the mean, as it is weighed heavily by extreme values. In these cases, the median would be a better way to look at central tendency.

Median
Median is the middle value of a data set. Since it is the middlemost item in a sample, half the numbers lie on each side of it. The median is used to represent central tendency, in cases where the mean could excessively skew results. For example, we commonly hear about the median income of a country. If we were to use the mean income, then high earners would skew the number excessively to one side, and this would not represent what most people earn. So, the median is more representative in this case.

Example:
Find the median of the following data set:
2,2,4,5,5,5,6,7,9,10,15
Here, the middle number is 5, so it is the median. There are 5 numbers on either side of 5, making it the median.

To find the median of an even set of numbers, the mean of the two centermost numbers is found, and designated as the median.

Example:
Find the median of the following data set:
2,4,6,6,8,9,12,15
Here, there are two middle numbers which are 6 and 8.
We take the mean of 6 and 8, which is 7. This is the median.

The median may also be a better indicator of central tendency when a set of numbers has an outlier. An outlier is an extreme value that differs greatly from the other values.

Mode
The mode is the most frequently occurring number in a data set. It is not influenced by numbers which may be very far from the center of a distribution. Sometimes a distribution has more than one mode. This could happen when the distribution represents the data from two processes.

Example:
Find the mode for the following data set:
2,3,3,5,5,5,6,7,16
Here, we see that 5 repeats 3 times, This repetition is more than any other number. So, 5 is the mode.

Measures of dispersion

There are three main measures of dispersion: Range, Variance, and Standard Deviation.

Range

Range is the difference between the largest and smallest values in a data set. It is affected only by the difference between the largest and smallest values in a given set of numbers.

Range R = largest number – smallest number

Example:

Find the range of the following data set:
1,2,3,3,3,4,6,8,10,10
Here, the largest number is 10, and the smallest number is 1.
So, the range is 10-1=9

Variance

Conceptually, variance of a population is the spread of data points from the mean. Mathematically, it is the average squared deviation from the population mean, as defined by the following formula:

$$\sigma^2 = \Sigma \, (\, X_i - \mu \,)^2 \, / \, N$$

where σ^2 is the population variance, μ is the population mean, X_i is each element of the population, and N is the total number of elements in the population.

The variance of a sample, is defined by slightly different formula, and uses a slightly different notation:

$$s^2 = \Sigma\,(\,x_i - x\,)^2 / (\,n - 1\,)$$

where s^2 is the sample variance, x is the sample mean, x_i is the ith element from the sample, and n is the number of elements in the sample. Using this formula, the sample variance is an unbiased estimate of the true population variance. This is the formula to use when estimating an unknown population variance, based on data from a sample.

Example:
A population consists of four observations: {1, 3, 5, 7}. What is the variance?
In Excel, select Formulas, More Functions, Statistical, VAR. Select the cells with the numbers identified above. The variance will show as 6.66

Standard Deviation
Standard deviation describes the dispersion of data within a distribution. Like variance, there are usually two standard deviations. One is the standard deviation of the population, and the other is the standard deviation of a sample of data from that population. If the standard deviation of the population is not known, then its value is estimated from a sample. It is referred to with the symbol σ, or sigma.

The standard deviation is the square root of the variance. Thus, the standard deviation of a population is:

$\sigma = \text{sqrt} [\sigma^2] = \text{sqrt} [\Sigma (X_i - \mu)^2 / N]$

where σ is the population standard deviation, σ^2 is the population variance, μ is the population mean, X_i is the ith element from the population, and N is the number of elements in the population.

The standard deviation of a sample is:
$s = \text{sqrt} [s^2] = \text{sqrt} [\Sigma (x_i - x)^2 / (n - 1)]$

where s is the sample standard deviation, s^2 is the sample variance, x is the sample mean, x_i is the ith element from the sample, and n is the number of elements in the sample.

The population standard deviation is usually unavailable as it is generally not possible to measure the entire population. Thus N, the number of values in the population, is not known, nor is μ, the population average.

The standard deviation for a population can be estimated from sample data. The sample average is substituted for the population average, and the sum of squared deviations in the sample is divided by the number of values in the sample, minus one, which biases this estimate of the population standard deviation on the higher side. This is especially important with small sample sizes of less than 30 values.

Example:
A sample consists of four observations: {1, 3, 5, 7}. What is the standard deviation?

In Excel, select Formulas, More Functions, Statistical, STDEV. Select the cells with the numbers identified above. The standard deviation will show as 2.58

Basic Distribution Shapes

Distributions of data could take various shapes, and identifying their pattern could form a basic diagnostic tool. If the distribution of the sample does match what is expected based on previous knowledge, then a 'special cause' could be driving this difference, and would warrant an investigation. Here are some of the basic shapes to recognize:

The Normal Distribution

There are many populations in nature, whose data is a 'normal' distribution. The meaning of a normal distribution is when most of the individual data points are clustered around the mean and a smaller number are spread on either side of the mean. This makes a normal distribution symmetrical in shape. This distribution is also referred to as the 'bell shaped' distribution. In the case of a perfectly normal distribution, the mean, median, and mode are in the center, and are the same.

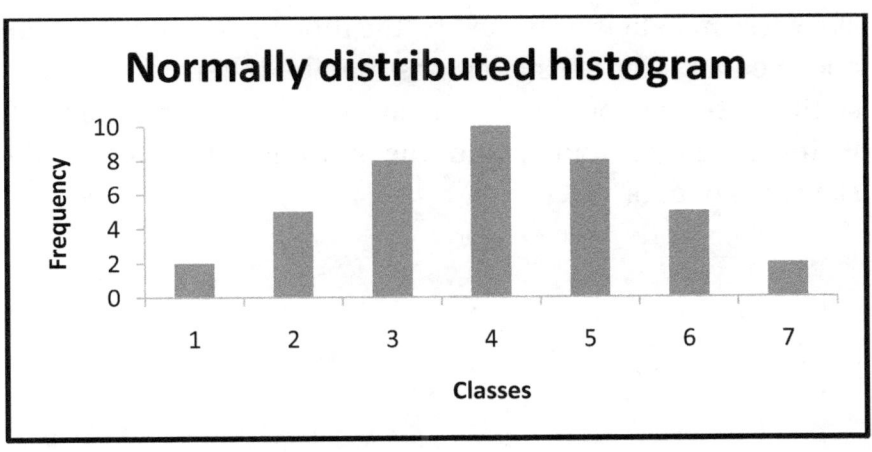

The Central Limit Theorem
'The distribution of an average tends to be normal, even when the distribution from which the average is computed is decidedly non-normal.'

The central limit theorem addresses an interesting idea which is used in control charts. Even if the population distribution under study is not normal, the sampling of the averages approaches a normal distribution. So, plotting the averages will allow one to use control charts.

The Skewed Distribution
When a distribution is not normal, it may be skewed. Data sometimes forms patterns which is not a symmetrical bell shape. The pattern may be skewed to one side or the other. Skewed distributions can be either positively or negatively skewed. In a positively skewed distribution, the tail of the distribution trails to the right. In a negatively skewed distribution, the tail of the distribution trails to the left. The skewness in data is indicated by the difference between mean and median. Some distributions may be inherently skewed, so this pattern should not be an automatic cause for alarm. It is important to understand the reasons which drive the skewness of data.

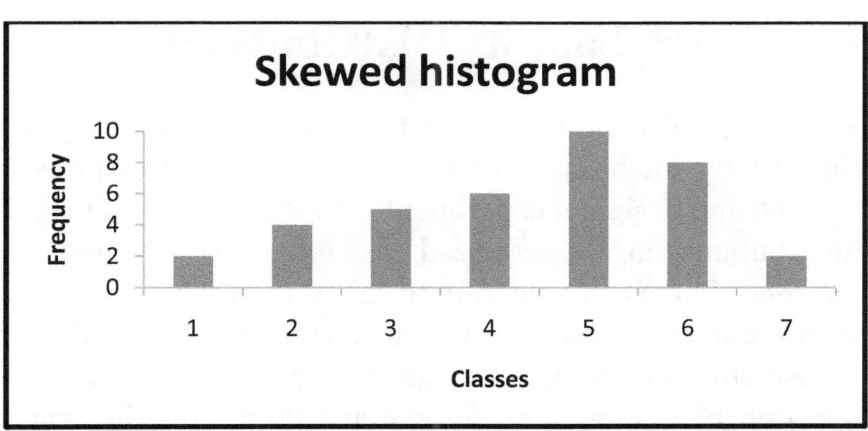

The Bimodal Distribution

In a bimodal distribution, the curve peaks twice.

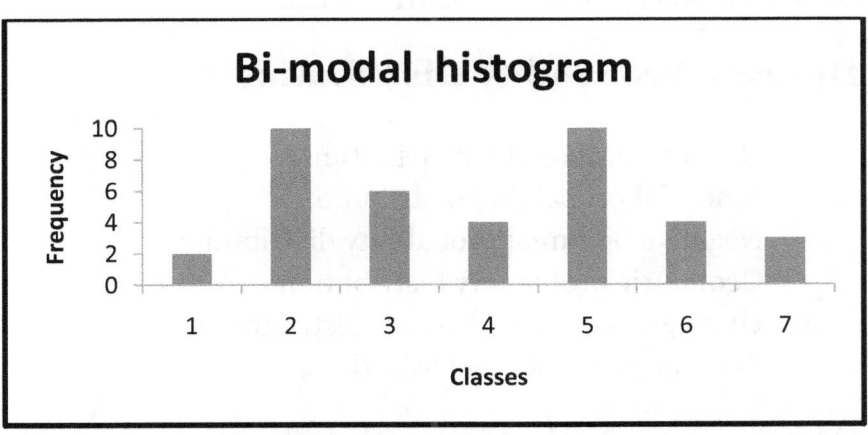

Probability Distributions

Probability distributions form the foundation of many six sigma tools such as hypothesis testing, statistical process control, and design of experiments. To start with, probability distributions are categorized as either continuous, or discrete. If data can take infinite values then the distribution is continuous. If data can take only few limited values, then it is discrete. For example, in the tossing of a coin, data can take only two values, a heads or a tails. So this distribution is discrete. On the other hand, in measuring the height of a group of people, data can assume any value, so the distribution is continuous. In general, if a distribution can take any value between two specified numbers, it is called a continuous distribution; otherwise, it is called a discrete distribution. Here are some of the commonly used discrete and continuous probability distributions.

Discrete Probability Distributions

- Uniform probability distribution
- Binomial probability distribution
- Negative binomial probability distribution
- Geometric probability distribution
- Hypergeometric probability distribution
- Poisson probability distribution

Continuous Probability Distributions

- Normal probability distribution

The Uniform probability distribution is unique since may take the form of a continuous or a discrete distribution, depending on the data source. It is reviewed as a discrete distribution here, but the same concept applies in the continuous situation.

Uniform Distribution

This distribution is seen when there is an equal chance of occurrence for all values of a random event. If a random variable X can assume k different values, then probability of X is one divided by k.

$$P(X) = 1/k$$

Example 1

Toss a die. What is the probability of the die landing on 4?
When a die is tossed, there are 6 possible outcomes represented by {1, 2, 3, 4, 5, 6}. Each possible outcome is a random variable (X), and each outcome is equally likely to happen. So this is a uniform distribution and $P(X = 4) = 1/6$

Example 2

In the same die tossing example, what is the probability that the die will land on a number that is smaller than 4?
This involves the calculation of cumulative probability. The probability that the die will land on a number smaller than 4 is equal to:

$$P(X < 4) = P(X = 1) + P(X = 2) + P(X = 3) = 1/6 + 1/6 = 1/2$$

The probability that the die will land on a number smaller than 4 is $1/2$.

Binomial Distribution

A binomial distribution is described as follows: Flip a coin three times and count the number of times the coin lands on tails. We can glean the main features of a Binomial distribution from this experiment.

Main features:
- There are repeated trials since the coin is flipped 3 times
- Each trial can result in only two possible outcomes - heads or tails
- The probability of success is constant - half on every trial
- The trials are independent; that is, getting heads on one trial does not affect whether we get heads on other trials

Example

A die is tossed 5 times. What is the probability of getting exactly 2 fours?

In this Binomial experiment, the number of trials is equal to 5, the number of successes is equal to 2, and the probability of success on a single trial is 1/6 or about 0.167. Entering these values in Excel gives us the binomial probability (formulas, more functions, statistical, binomdist)

$b(2; 5, 0.167, false) = 0.161$

where
number of successes = 2
number of trials = 5

Probability of success in each trial = 0.167
Cumulative = false

So the probability of getting exactly two fours is in five throws is 0.16

Negative Binomial Distribution

A negative binomial distribution is described as follows: Flip a coin repeatedly and count the number of times it lands on heads. Continue flipping the coin until it lands 5 times on heads. We can glean the main features of a negative Binomial distribution from this experiment.

Main features:

- The experiment consists of repeated trials. A coin is repeatedly flipped until it has landed 5 times on heads
- Each trial can result in only two possible outcomes - heads or tails
- The probability of success is constant - half on every trial
- The trials are independent; that is, getting heads on one trial does not affect whether we get heads on other trials
- The experiment continues until a fixed number of successes have occurred; in this case 5 heads

Example

Bob is a high school basketball player. He is a 70% free throw shooter. That means his probability of making a free throw is 0.70. During the season, what is the probability that Bob makes his third free throw on his fifth shot?

In this negative binomial experiment, the probability of success (P) is 0.70, the number of trials (x) is 5, and the number of successes (r) is 3. Entering these values in Excel gives us the negative binomial probability (formulas, more functions, statistical, negbinomdist)

nb(2; 3, 0.7) = 0.18522
where, number of failures = 2
number of trials = 3
Probability of success in each trial = 0.7

Thus, the probability that Bob will make his third successful free throw on his fifth shot is 0.18522

Geometric Distribution
The geometric distribution is a special case of the negative binomial distribution. It shows the number of trials required for a single success. Thus, the geometric distribution is negative binomial distribution where the number of successes (r) is equal to 1

Example
From the previous example, a slightly different question can be asked: What is the probability that Bob makes his first free throw on his fifth shot?

This is a geometric distribution, which is a special case of a negative binomial distribution. So, it can be solved using the negative binomial formula.

The probability of success (P) is 0.70, the number of trials (x) is 5, and the number of successes (r) is 1. Entering these values in Excel gives us the negative binomial probability (formulas, more functions, statistical, negbinomdist)

nb(4; 1, 0.7) = 0.00567
where
number of failures = 4
number of trials = 1
Probability of success in each trial = 0.7

Thus, the probability that Bob will make his first free throw on his fifth shot is 0.00567

Hypergeometric Distribution

For a hypergeometric distribution, consider the following experiment. There is a bag of 10 marbles - 5 red and 5 green. Randomly select 2 marbles without replacement and count the number of red marbles selected. This is not a Binomial experiment since it requires the probability of success to be constant on every trial. Here, the probability of a success changes on every trial. At first, the probability of selecting a red marble is 5/10. If red marble is selected on the first trial, the probability of selecting a red marble on the second trial is 4/9. If a green marble is selected on the first trial, the probability of selecting a red marble on the second trial is 5/9. If marbles are selected with replacement, the probability of success will not change. It would be 5/10 on every trial and be a Binomial experiment.

Example

Randomly select 5 cards without replacement from an ordinary deck of playing cards. What is the probability of getting exactly 2 red cards (i.e., hearts or diamonds)?

This is a hypergeometric experiment in which we know that
- $N = 52$; since there are 52 cards in a deck.
- $k = 26$; since there are 26 red cards in a deck.
- $n = 5$; since we randomly select 5 cards from the deck.
- $x = 2$; since 2 of the cards we select are red.

Entering these values in Excel gives us the hypergeometric probability (formulas, more functions, statistical, hygeomdist)

hg(2, 5, 26, 52) = 0.32513
where, sample size = 2
number of samples = 5
success in the population = 26
population size = 52

Thus, the probability of randomly selecting 2 red cards is
0.32513

Poisson Distribution

A Poisson distribution is used in scenarios in which the outcomes can be classified as successes or failures, the average number of successes is known, and the probability of exactly x successes needs to be determined.

Example

The average number of homes sold by the Hakim realty company is 2 homes per day. What is the probability that exactly 3 homes will be sold tomorrow?

This is a Poisson experiment in which we know the following:

- $\mu = 2$; since 2 homes are sold per day on average.
- $x = 3$; since we want to find the likelihood that 3 homes will be sold tomorrow.

Entering these values in Excel gives us the Poisson probability (formulas, more functions, statistical, poisson)

p(3, 2, false) = 0.180

where
number of events = 3
mean = 2
logical value = false

Thus, the probability of selling 3 homes tomorrow is 0.180

	Explanation	Required Information	Excel Function
Binomial	n trials result in x successes. The probability of success on an individual trial is P	number of successes x, number of trials n probability of success in each trial P, logical value = false	binomdist
Negative Binomial	x trials result in r successes. The probability of success on an individual trial is P	number of failures x-r, number of trials x, Probability of success in each trial P	negbinom dist
Geometric	x trials result in one success. The probability of success on an individual trial is P	Same as negative binomial	negbinom dist
Hypergeo metric	N items, k of which are successes. A random sample drawn from that population	sample size N number of samples k, success in the population x, population size n	Hygeom dist

	consists on n items, x of which are successes		
Poisson	x is the actual number of successes that result from the experiment, and e is approximately equal to 2.71828	number of events x, mean μ, logical value = false	poisson

Normal Distribution

A normal distribution is also referred to as a 'bell shaped' curve. The normal distribution is built around two measures; mean and standard deviation. The mean determines the center of the distribution, and the standard deviation determines its height and width. When the standard deviation is large, the curve is short and wide; when it is small, the curve is tall and narrow.

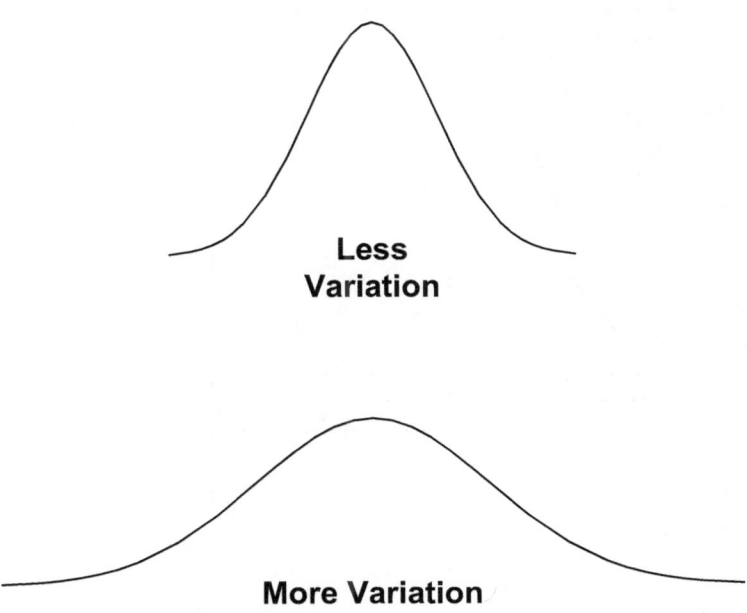

Less Variation

More Variation

The curve at the bottom is shorter and wider than the curve on top because it has a larger standard deviation.

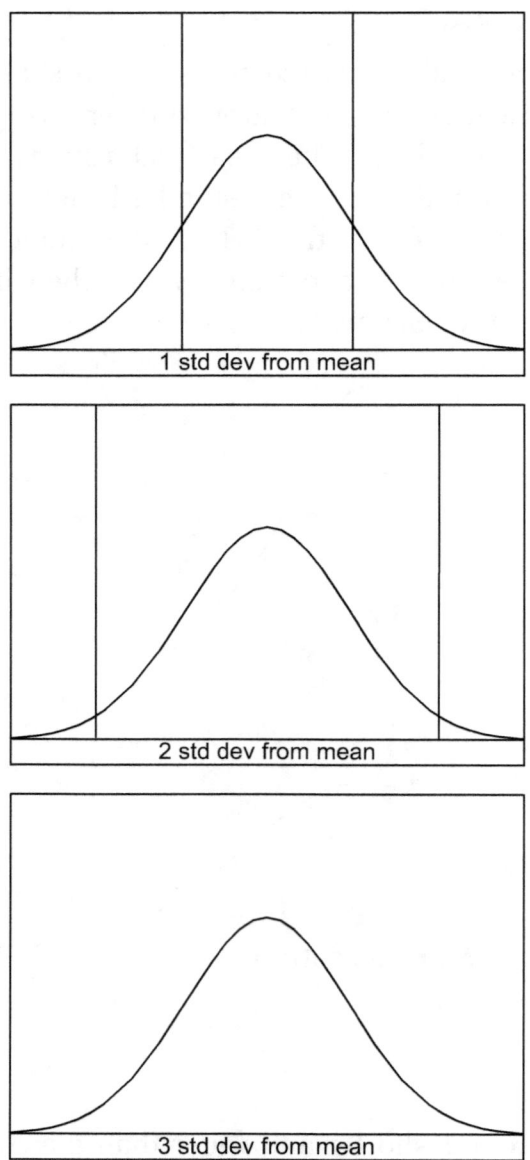

For all normal curves, about 68% of the area falls within 1 standard deviation of the mean, 95% falls within 2 standard deviations of the mean, and 99.7% falls within 3 standard deviations of the mean.

For a normal distribution, most outcomes will lie within 3 standard deviations of the mean. We can continue with this thought and move further away from the mean to the fourth, fifth, and sixth standard deviations. By the time we reach the sixth, we cover 99.9996599% of the normal curve. This is where the term six sigma comes from. It refers to ensuring that six standard deviations of products, processes, or services are within required specifications. This is its goal.

Chapter 8: Making it Happen

At 7:00 a.m., as Dean walked in he saw Jeff in the office and murmured, "Working hard, Jeff?" At this point, these words did not mean much. Jeff quickly got to his feet and said, "Dean, you have to see what I have written up in the past few days. I need some of your time right now, since I think it is very important for us to review this."

Jeff proceeded to show Dean his writings. "Some of this does sound very familiar to me," Dean began to say. "That's because they are your words," said Jeff, "and some of them are my own, what I have learned from my mistakes and my experiences." Dean took a lot longer in reviewing his notes than Jeff had expected. "The boss is definitely interested, but I hope he does not find issues with what I have written. However, if he does, I have reached a point where I can confidently challenge what he says. I know what I am talking about," Jeff thought. What Jeff did not anticipate was that Dean was thinking ahead, not backwards. "Jeff," said Dean, "What you have done is created an efficient way to manage your business. What this does is free up your time to do what you are really paid to do. As a manager, you are given resources to get things done. You have to manage your resources and leverage them in order to accomplish your greater goal. Since you have done such a good job in documenting these best practices and developing an action items list, I would suggest that you close the loop and add to it the most important piece of the puzzle."

This was something that Jeff did not understand. His mind was racing, standing there in front of Dean, as he tried to buy seconds. "What does he mean by the most important piece of the puzzle? Could it be the details of the plan, the charts, reports with actual data,...? Or, on a different note, could it have to do with creating new processes, and delivering results. Yes, of course. That is the most important part of the puzzle, to innovate and develop."

"You are right, Dean," said Jeff after a pause. "I do need to add the most important piece to this plan, and that is how to use this system to make things happen in our organization. So far, what I have is benign in nature. I need to take this, and use it as a springboard to determine next steps, to innovate new solutions, and challenge the very foundation of some of the things we do. The thing is that I can do this now, with all the documents, processes, systems, knowledge, and data available on hand."

After Dean left, Jeff wrote down the most important step of the process:

Innovate and develop: Once all the previous steps have been addressed, this should free up a chunk of your time that you used for other mundane activities. This time can now be funneled to develop processes, and create new ideas. Identifying opportunities for improvements is a disciplined process. Ask the following questions to find these opportunities:

  a. What are the current processes we use? How can we streamline these processes?

b. What else could we do to improve the value of our products/services?
c. How can we solve problems that constitute the biggest percentage of our business?
d. What is our vision of the ideal product or service?

Desired State

Each day, after spending time on 'essential' work, get creative with innovating solutions that will improve customer satisfaction, reduce costs, increase revenue, achieve business goals, or improve productivity. This is the kind of stuff that you will enjoy working on, and will help you achieve the success you desire.

Six sigma quality has tools which will help innovate and develop new processes, products, or services. By themselves, they do not help the actual innovation or development. What they do is determine if the new process, product, or service is truly better. They are the catalysts for improvement. These tools remove the emotion out of finding out if things have improved. They prove it with numbers. Hypothesis testing, analysis of variance, and regression analysis are the tools to learn here. The selection of a particular tool depends on what is to be tested. Finally, measurement analysis ensures that one's measurements are viewed with the right lens and results are not taken out of context.

Hypothesis Testing

Hypothesis testing is used to determine if there is a significant difference between two groups. To use this tool, we define a 'null hypothesis' represented as H_0 and an 'alternative hypothesis' represented as H_1. The null hypothesis reflects no difference between groups, while the alternative hypothesis points to a significant difference between them.

Terminology used in hypothesis testing:

Degrees of freedom: This is the amount of freedom we have to change various values in an equation. For example if we have three numbers that add up to 60, then the first two numbers can be a whole list of numbers, but when we get to the third number we do not have any freedom if the sum is to be 60. So, with three numbers we have two degrees of freedom.

p-value is a measure of how much evidence we have against the null hypothesis. The null hypothesis H_0 represents no change or no effect. The smaller the p-value, the more evidence we have against H_0.

Procedure for using hypothesis testing

1. Establish the hypothesis

Null hypothesis - difference between observed and expected is zero

Alternative hypothesis - difference between observed and expected is not zero

2. Calculate the test statistic

This calculation depends on the type of test being done.

3. Determine p-value

After calculating the t-statistic, compare the t-value with a standard table of t-values to determine whether it reaches the threshold of statistical significance.

A p-value of .05 or less is required to reject the null hypothesis.

Errors in hypothesis testing:

Type one error: A 'type one error' is rejecting the null hypothesis when it is true. This means that there was no change between two groups, but we interpret that there was a change. The probability of making a type one error is called alpha. We want this value to be small, so we should plan an experiment in which alpha is either 0.05 or even 0.01

Type two error: A 'type two error' is accepting the null hypothesis when it is false. This means that there was a change between the two groups, but we interpret that there was no change. The probability of making a type two error is called beta.

Types of hypothesis tests:

The following are some of the commonly used hypothesis tests:

- Z test
- One sample t-test
- Two sample t-test
- Paired t-test
- Chi-square test

Test Name	This is used when...
Z test	comparing a sample mean with a population mean and the standard deviation for the population is known
One sample t-test	comparing a sample mean with a population mean, and the standard deviation of the population is not known
Two sample t-test	comparing the difference between the means of two groups and the two groups are **independent** of each other, each having an underlying normal distribution
Paired t-test	comparing the difference between the means of two groups and the two groups are **not independent** of each other, and each has an underlying normal distribution
Chi-square test	comparing the **observed frequency** of some observation (such as frequency of buying different brands of computers) with an **expected frequency** (such as buying equal numbers of each brand of computer)

Z test

This is used when comparing a single group's mean with a population mean and the standard deviation for the population is known.

In many cases, the mean and standard deviation of populations are known. For example, the scores of all students taking certain standardized tests are recorded, so the mean and standard deviation of the entire population of students who have taken these tests thus far is known. In this scenario, it is possible to compare the performance of an individual, or a group of individuals to the overall population. A good way to do this comparison is with the z-test.

Standard scores (z-Scores)

The z-score will indicate how many standard deviations away the sample is from the population mean. To conceptually understand the z-score, think of the normal distribution with the mean line at the center. If the z-score is zero, then the sample mean is the same as the population mean. A positive z-score indicates the sample mean is larger than the population mean, while a negative z-score shows that it is less than the population mean. A z-score between -1 and +1 shows the sample to be within one standard deviation of the population mean. Similarly, z-scores over 2, but less than 3 indicate the sample to be over 2 standard deviations from the population mean, while z-scores in the threes indicate the sample to be significantly away from the population mean. For normal distributions, about 68% of data points have a z-score between -1 and 1; about 95% have a z-score between -2 and 2; and about 99% have a z-score between -3 and 3

The z-score can be calculated with the following formula

$$z = (X - \mu) / \sigma$$

where z is the z-score, X is the value of the sample, μ is the mean of the population, and σ is the standard deviation of the population.

One sample t-test

This test is used to compare a sample mean with a population mean, when the population standard deviation is not known. This occurs in many real life scenarios, where it is not possible to determine the standard deviation of the population. For example, consider a scenario in which we wanted to see if the effects of a pre-natal care program had a positive impact on the health of a child, as measured by its weight at birth. In this example, it may not be possible to know the standard deviation of the weight of all children who were not a part of this program.

Steps:
1. Establish hypothesis
Determine the measurement you want to compare. Make sure you have the sample data.
Null hypothesis – there is no difference between sample and population data.
Alternative hypothesis– there is a difference between sample and population data.

2. Calculate test statistic
Calculation of the test statistic requires four components:
    A. The average of the sample (observed average)
    B. The population average (expected average)
    C. The standard deviation of the sample
    D. The number of observations

3. Determine p-value
Once the test statistic is calculated, compare it with a standard table to determine whether the p-value reaches the

threshold of statistical significance. We require p-values of .05 or less to reject the null hypothesis.

Two sample t-test

This test is used to compare the difference between means of two groups. These two groups need to be independent of each other, each having an underlying normal distribution.

1. Establish hypothesis

Determine the measurement you want to compare. Make sure you have the sample data.

Null hypothesis – there is no difference between sample and population data.

Alternative hypothesis– there is a difference between sample and population data.

Here is a sample of two data sets we may want to compare

1	2
5	4
4	6
3	8
6	1
2	8

In Excel, go to Data, Data Analysis, t-Test: Two sample assuming equal variances.

2. Calculate test statistic

Select the first data set in the variable 1 range. Select the second data set in the variable 2 range. Select an alpha level, or keep the default of 0.05. Select OK. You will see a new worksheet with the t-test calculations appear. Observe the mean and variance of the two samples. Then look at the degrees of freedom (df) cell. The df will be the sum of the two sample sizes, minus two. Now look at the t-statistic, and then the p value for the one tailed test.

	Variable 1	Variable 2
Mean	3.5	4.833333
Variance	3.5	8.966667
Observations	6	6
Pooled Var	6.233333333	
H_0	0	
df	10	
t-stat	-0.92499458	
p one-tail	0.188375292	
t critical 1 tail	1.812461102	
p two tail	0.376750585	
t critical 2 tail	2.228138842	

3. Determine p-value

We require p-values of .05 or less to reject the null hypothesis. In the above example, p-value for the one tailed test is 0.188, which is greater than 0.05. So, we cannot reject the null hypothesis, meaning cannot say that there is a significant difference between the two means.

Paired t-test

A 'paired t-test' is used when we want to see if there is a significant difference between the means of two samples, but the two samples are not independent of each other. There are two situations in which the two samples may not be independent:

1. When two samples are matched before being put in the two groups. For example, consider an experiment measuring the effect of a new method of teaching on comprehension test scores. If we want to eliminate the effect of intelligence in the two groups, we can distribute students evenly to remove this bias. Prior to the grouping, we give an intelligence test to all students in the study. Then we take the two students with the highest IQ and randomly put one of them in the experimental group and the other in the control group. We then take the two students with the next highest intelligence and do the same thing until we have selected our two groups but they are matched on the factor of tested intelligence.

2. When two groups are the same subjects being administered a pre-test and a post-test. For example, consider the study to test the effect of a new teaching method on comprehension scores. We could administer a comprehension test to the group as a pre-test, and then apply the experimental teaching method to the group of students. This can be followed up by a post-test. We then see if there is a significant difference between the two scores.

When subjects are connected to each other by either of these methods we use a test to measure the significance of difference between the means. We must use a test that takes into consideration these dependencies. This is what the dependent or paired t-test does.

Consider the following data set of scores from the pre-test and post-test.

65	69
68	76
73	78
90	96
25	60
64	68

Establish hypothesis
Null hypothesis – there is no difference between the two sample data sets.
Alternative hypothesis– there is a difference between the two sample data sets.
In Excel, go to Data, Data Analysis, t-Test: paired sample.

2. Calculate test statistic
Select the first data set in the 'variable 1 range'. Select the second data set in the 'variable 2 range'. Select an alpha level, or keep the default of 0.05. Select OK. You will see a new worksheet with the paired t-test calculations appear. Observe the mean and variance of the two samples. Then look at the degrees of freedom (df) cell. The df will be the sample sizes of one column, minus one. Now look at the t-statistic, and then the p value for the one tailed test.

	Variable 1	Variable 2
Mean	64.16666667	74.5
Variance	458.9666667	151.9
Observations	6	6
Pooled Var	0.876004169	
H_o	0	
df	5	
t-stat	-2.078711901	
p one-tail	0.04610637	
t critical 1 tail	2.015048372	
p two tail	0.092212739	
t critical 2 tail	2.570581835	

3. Determine p-value

We require p-values of .05 or less in order to reject the null hypothesis. In the above example, p-value for the one tailed test is 0.046, which is less than 0.05. So, we can reject the null hypothesis, meaning that there is a significant difference between the two means.

Chi square test

The chi-square test is a statistical test used to examine differences with categorical variables. It is used to compare the observed frequency (such as frequency of buying different brands of computers) with an expected frequency (such as buying equal numbers of each brand of computer). The comparison of observed and expected frequencies is used to calculate the value of the chi-square statistic. The chi-square test is used in two similar but distinct circumstances:

    a. Estimating how closely an observed distribution matches an expected distribution. This is also called the goodness-of-fit test

    b. Estimating whether two random variables are independent

The symbol for chi-square and the formula are as follows:

$$\sum \frac{(O - E)^2}{E}$$

where

O is the observed frequency, and

E is the expected frequency.

The degrees of freedom df is: C - 1

where C is the number of categories or levels of the independent variable

Let's look at an example in which a person plans to sell T-shirts at a country fair. She estimates that of the 5 colors she has, all will be in equal demand, so each will have an expected frequency of 20%. She finds that during the fair, people did not select all colors equally. She made a note of what was sold, and documented that as the observed frequency. She now wants to determine if there was a significant difference between color preferences, or it was just normal variation. Using Excel to perform calculations, prepare data for the observed and expected frequencies:

	O	E
Red	14	20
Blue	18	20
Green	24	20
Yellow	26	20
White	18	20

In Excel, go to formulas, more functions, statistical, chitest. In the 'actual' range, select the first column numbers and in the 'expected' range select the second column numbers. Click OK. We obtain a resulting p-value of 0.3. Since this is not less than 0.05, we conclude that there was no significant difference between the observed and expected preferences for buying the T-shirts.

Anova (F-test)

The t-test is a useful tool to compare the means of two groups but is not good for comparing three or more groups. It can only compare one group's mean to a known distribution or compare the means of two groups. There are many situations in which we may want to compare means of many different groups. In these scenarios, the preferred statistical tool is ANOVA, or Analysis Of Variance. ANOVA produces an F-statistic which indicates if there is a significant difference among the sample means.

ANOVA uses variance to compare multiple averages at the same time. Instead of comparing pair wise averages, it compares the variance between groups to the variance within groups. If the variance between groups is the same as the variance within groups, there is no difference between the group averages.

ANOVA Steps

1. Calculate the variation between groups.
2. Calculate the variation within groups. If they are about the same, there is no significant difference among groups.
3. Calculate the ratio of two variances. This is the F-statistic.
4. Get a p-value from the F-distribution. We then test the significance of F to complete our analysis of variance.

ANOVA Assumptions

Analysis of Variance methods have two assumptions:
1. The standard deviations of the populations for all groups are equal, and
2. The samples are randomly selected from the population

Explanation for performing ANOVA:
1. Calculate variation between groups
The first step is to calculate variation between groups by comparing the mean of each group with the mean of the overall sample. This measure of between-group variance is referred to as 'between sum of squares' or BSS. The sum of squares has degrees of freedom equal to the number of groups minus 1. If there are three groups, then $df_B = (3\text{-}1) = 2$

Divide the BSS figure by the degrees of freedom to get the variation between groups, referred to as 'Between Mean Squares'.

2. Calculate variation within groups
To measure variation within groups, find sum of the squared deviation. This sum is referred to as the 'within sum of squares' or WSS.

Take a value equal to the number of cases in the total sample minus the number of groups to get dfw. Divide WSS by the degrees of freedom to get the variation within groups, referred to as 'Within Mean Squares'.

3. Calculate the F-test statistic
For this calculation, divide the Between Mean Squares, the value obtained in step 1, by the Within Mean Squares, the value calculated in step 2.

Compare this value to a standard table with values for the F distribution to calculate the significance level for the F value. If the significance level is less than .05, then there is strong evidence against the null hypothesis.

Here is a sample ANOVA table:

Source	Sum of Squares	Degrees of Freedom	Mean Squares
Between	BSS	df_B	Between Mean Squares BSS/df_B
Within	WSS	df_W	Within Mean Squares WSS/df_W
Total	TSS = BSS + WSS		

Consider an example with the following data set representing three groups. We want to determine if there is a significant difference among these three groups.

12	14	20
15	13	18
16	16	15
18	19	19
20	10	11
14	10	12

In Excel, go to Data, Data Analysis, analysis of variance, one factor.

Select the entire data set in the variable range. Keep the default alpha level of 0.05. Select OK. You will see a new worksheet with the ANOVA calculations. Observe the summary of the three groups, and their corresponding averages and variances.

Summary

Groups	Count	Sum	Average	Variance
Column 1	6	95	15.83333	8.166667
Column 2	6	82	13.66667	12.26667
Column 3	6	95	15.83333	14.16667

Now look at the ANOVA calculations. We see a p-value of 0.46. This suggests that there is not enough evidence to claim that there is a significant difference among the groups.

Analysis of Variance

Source of Variation	SS	df	MS	F	P value
Between groups	18.77778	2	9.388889	0.814066	0.461699
Within groups	173	15	11.53333		
Total	191.7778	17			

Regression Analysis

The purpose of regression is to learn more about the relationship between several independent variables, and a dependent variable. It allows us to look at the effects of many factors on some outcome. This relationship can be observed on a scatterplot, which will show the correlation between the factor and the outcome. The scatterplots below show degrees of correlation for different patterns of data.

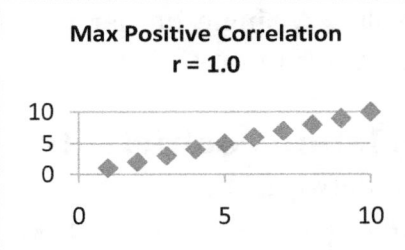

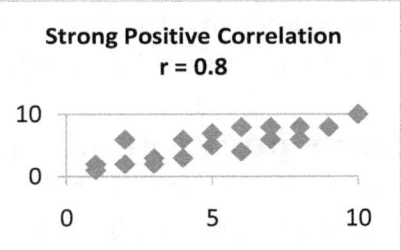

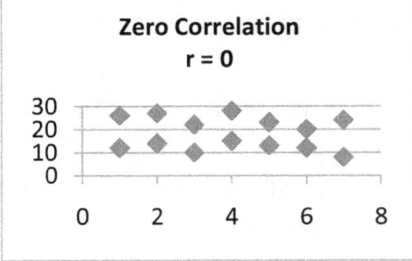

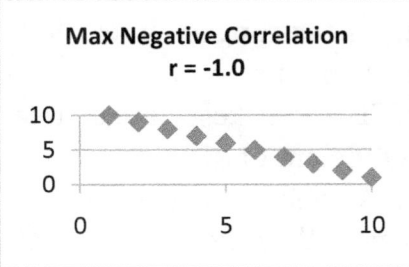

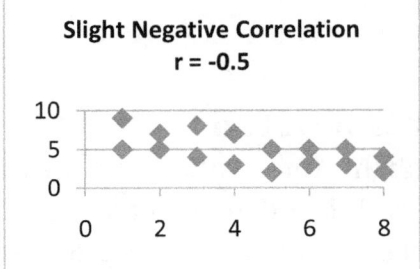

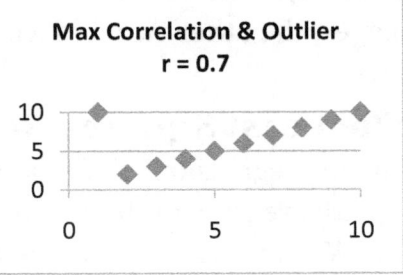

In the preceding scatterplots, r is the correlation co-efficient. When the slope of the line in the plot is negative, the correlation is negative, and vice versa. The strongest correlations (r = 1.0 and r = -1.0) occur when data points fall exactly on a straight line. Correlation becomes weaker as the data points become scattered. If data points fall in a random pattern, the correlation is equal to zero.

Correlation is affected by outliers. Compare the first scatterplot with the last scatterplot. A single outlier in the last plot greatly reduces correlation from 1.00 to 0.7

Some points about correlation coefficients
1. A correlation coefficient ranges between -1 and 1
2. The greater the absolute value of a correlation coefficient, the stronger the linear relationship between two factors
3. The strongest linear relationship is with a correlation coefficient of -1 or 1
4. The weakest linear relationship is with a correlation coefficient of 0
5. A positive correlation means that if one variable gets bigger, the other variable will also get bigger
6. A negative correlation means that if one variable gets bigger, the other variable will get smaller

The Least Squares Regression Line
Linear regression finds a straight line, called the 'least squares regression line'. Suppose Y is a dependent variable, and X is an independent variable. The population regression line is:

$Y = B_0 + B_1X$

where B_0 is a constant, B_1 is the regression coefficient, X is the value of the independent variable, and Y is the value of the dependent variable. The dependent variable Y has a linear relationship to the independent variable X.

How to Define a Regression Line

We will use a computational tool in Excel to find b_0 and b_1. We enter the X and Y values, and the tool solves for each parameter. The regression analysis tool performs linear regression analysis by using the 'least squares' method to fit a line through a set of observations. You can analyze how a single dependent variable is affected by the values of one or more independent variables.

The Coefficient of Determination or R^2

The coefficient of determination is denoted by R^2 and is a key output of regression analysis. It is interpreted as the proportion of the variance in the dependent variable that is predictable from the independent variable. R^2 ranges from 0 to 1. An R^2 of zero means that the dependent variable cannot be predicted from the independent variable. An R^2 of 1 means the dependent variable can be predicted without error from the independent variable. An R^2 between 0 and 1 indicates the extent to which the dependent variable is predictable. An R^2 value of 0.10 means that 10 percent of the Y variance is predictable from X. An R^2 of 0.20 means that 20 percent is predictable, and so on.

Standard Error

The standard error about the regression line is a measure of the average amount that the regression equation over or under predicts. Higher coefficient of determinations, point to lower standard errors, and more accurate the predictions.

The linear regression model is not always appropriate for all data. Appropriateness of the model can be assessed by examining residuals, outliers, and influential points.

Residuals

The difference between the observed value of the dependent variable (y) and the predicted value ($\hat{y}$) is called the residual (e). Each data point has one residual. Both the sum and the mean of residuals are equal to zero.

Residual = Observed value - Predicted value

$e = y - \hat{y}$

Residual Plots

A residual plot is a graph that shows residuals on the vertical axis and the independent variable on the horizontal axis. If the points in a residual plot are randomly dispersed around the horizontal axis, a linear regression model is appropriate for the data; otherwise, a non-linear model would be better.

The following chart displays a residual plot, which is a straight line. Since it is non-random, a non-linear model will provide a better fit to the data or it may be possible to "transform" the data to allow use of a linear model.

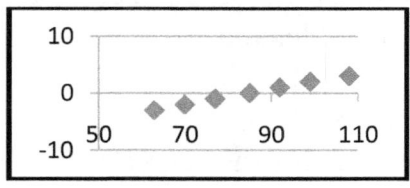

Below, the residual plots show typical patterns. The first plot shows a random pattern, indicating a good fit for a linear model. The other plot is non-random, suggesting a better fit for a non-linear model.

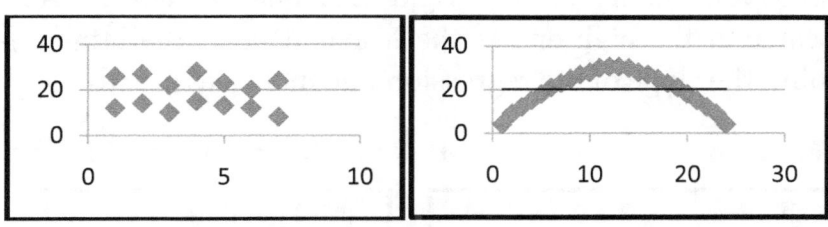

Random pattern   Non-random: Inverted U

Outliers

Data points that diverge from the overall pattern and have large residuals are called outliers. Outliers limit the fit of the regression equation to the data. This is illustrated in the scatterplots below. R^2 is bigger when outliers are not present.

Without Outlier   **With Outlier**

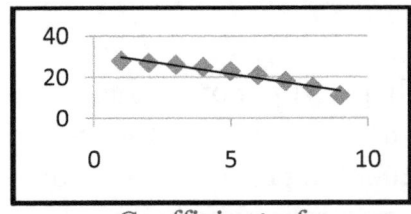 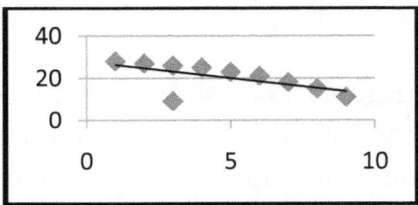

Coefficient of   Coefficient of determination:

determination: $R^2 = 0.95$ $R^2 = 0.6$

Outliers reduce the ability of a regression model to fit the data, and thus reduce the coefficient of determination.

Influential Points

Influential points are data points with extreme values that greatly affect the slope of a regression line. The charts below show regression for a data set with and without an influential point. The chart on the right has one influential point, located at the high end of the X axis. Due to the influential point, the slope of the regression line increases greatly.

Without Influential Point   **With Influential Point**

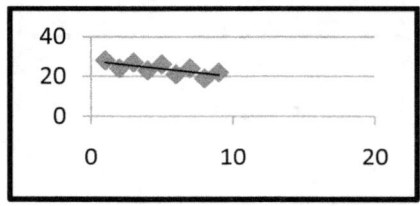   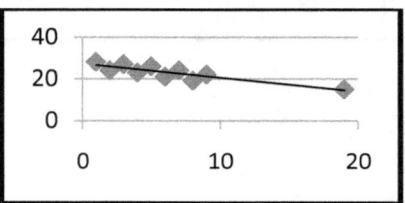

Regression equation: $\hat{y} =$

92.54 - 2.5x

Slope: $b_0 = -2.5$

Coefficient of determination:

$R^2 = 0.46$

Regression equation: $\hat{y} =$

87.59 - 1.6x

Slope: $b_0 = -1.6$

Coefficient of determination:

$R^2 = 0.52$

Influential points, unlike outliers, do not reduce the coefficient of determination. In fact, the coefficient of determination is bigger when influential points are present. Influential points increase the correlation coefficient.

Measurement Analysis

Measurement systems are used in a variety of industries, which include manufacturing, R&D, and marketing. Things that are measured include distance, temperature, strength, and sales, just to name a few.

Measurement instruments should not have an error greater than 10% of the measure being taken, for them to accurately portray the measurement. If the error is greater than 10%, then the readings observed are suspect. For example, if a part dimension, which is specified to be 3 inches is being measured, then the measuring instrument cannot have a measurement error that exceeds 0.3 inches, for the readings to be deemed valid.

The field of measurement involves several terms. Some of the commonly used ones are:

Gages: A gage is a measuring instrument

Gage R&R: Gage R&R stands for gage repeatability and reproducibility. It is a statistical tool that measures the amount of variation in the measurement system due to a combination of the measurement device and the people taking the measurement. Gage variation is called repeatability and user variation is called reproducibility.

Repeatability: Repeatability is the variation from one gage and one user when measuring the same part several times.

Reproducibility: Reproducibility is the difference in measurements from several users using the same gage and measuring the same part. It shows the user-to-user variation.

Linearity: Gage linearity is used to identify the accuracy of measurements through the entire range of measurements.

Accuracy: Accuracy is the closeness of measurements between the actual values measured, and the master value measured for that characteristic.

Bias: Bias is the numerical value used to measure accuracy. It is the difference between the average value of all the measurements and the master value.

Stability: Stability is the consistency of performance over time. It can be tracked using a control chart by measuring the same part over time.

Precision: Precision is the degree of repeatability between measurements. The smaller the spread of the measurements, the better the precision is.

Methods for Calculating Gage R&R

Gage R&R is done to determine repeatability and reproducibility of the measurement system. This evaluation should be done on a periodic basis to ensure that the gages are providing useful data.

In order to start a gage R&R study, perform the following steps:

1. Identify the measurement system is to be studied.

2. Ensure that the test procedure is well documented.

3. Decide on the number of people participating in the study (k), the number of sample parts (n), and the number of repeated readings (r). A common approach is to select 3 people, 4 to 10 sample parts, and 3 to 6 repeated readings. Increase in the number of people, parts, or readings, increases confidence of results. It also increases the cost and time for the study.

4. The people selected for the study should be trained in using the measurement system. Have the parts be available before the study begins. The sample parts must come from an existing process and represent its entire measuring range. Ensure that parts are randomly selected for the study.

5. Randomize the (n) samples and (k) people to determine the measurement order for the first run.

6. Repeat the randomization process for each of the following runs. The person leading the study should read and record data so that it remains hidden from those measuring it.

7. Use an appropriate gage R&R spreadsheet to perform calculations.

Chapter 9: Innovate and Develop

Jeff was happy that he understood Dean's train of thought and was able to answer his question in a matter-of-fact way. It was with a lot of self disciplined training that he had reached this stage. There was one thing though, which he thought Dean did not ask about, but should have. That was, "How are you going to accomplish the most important piece of the puzzle?" It is one thing to say, 'create systems that allow you to spend time to innovate and develop'. It is another thing to actually do it. What are the tools needed to accomplish this? How do you begin? How can one make it seem effortless? As these questions popped in his head, Jeff began to wonder if these were all manifestations of the same question, which basically had one answer. Giving it a bit more thought, Jeff realized that he was lacking some basic tools that would help him determine opportunities for improvements within his organization. He wanted to understand these tools, and even create new ones, if required. Skimming through various books, he found that there were some tools that he would be able to use. In other cases, his thought was to come up with some himself that would help him mange his business better. In order for others within the organization to be able to use the tools, he wanted them to be as generic as possible, so that they could be customized to meet the needs of each individual, in varied circumstances. One tool Jeff thought that would be great to have would be the 'balanced' scorecard. "This was something that Dean might need more than me," he thought.

Jeff was excited to learn something new that would allow him to multiply the effect of his efforts. Integrating statistics into common sense management would enable him to use his time in a more enlightened way, racing towards improving quality and productivity for the organization. All these goals could now be accomplished without getting stressed and wondering if things were moving in the right direction. Time spent at work would be something to look forward to. He would be able to make intelligent decision, and get creative juices flowing when working on projects.

Desired State: Statistical tools are used as a framework to spur innovation and development. Decisions are based on sound statistical principles and 'tells' from upper management are challenged with data.

There is a common misconception among organizations that use of statistical data is academic in nature and does not help solve real world problems. There is a grain of truth to this, since data is sometimes used in a theoretical sense, without associated actions. However, action backed by data is the most powerful method to create innovation and improve processes. Design of Experiments is a statistical tool to determine the best mix of variables to innovate new solutions. Statistical Process Control or SPC allows one to monitor the stability of a process and improve it by reducing 'common cause' variations. These tools should be first understood and then appropriately deployed within the fabric of an organization's work.

Design of Experiments

Design of Experiments (DOE) was first proposed in 1935 by Ronald Fisher in his book 'The Design of Experiments'. It is an organized approach to determine the impact of various factors, also known as input variables (Xs) on the output variable (Y). It identifies the sources of variation and determines which ones have the most impact on results. It also quantifies the effects of interactions of input variables on the output and produces an equation between Xs and Y. It makes use of existing product or process knowledge, by identifying a combination of the most appropriate factors to quickly get to an answer.

The following are some of the benefits of a good experimental design:
- Increased process output
- Reduced variation
- Reduced costs
- Reduced development and design time

There are different ways an experiment may be conducted. These methods are listed below from most simple to the most complex. It is also in the order of increasing value, and their ability to generate insights.

- **Trial and error experiment:** This introduces a change in a process to see what happens
- **One factor experiment:** This changes one factor at a time, while keeping all other factors constant

- **Experiment with two to four factors:** This studies the main effects of the factors and their interactions. These are mainly full factorial experiments, but could also be fractional factorial or Taguchi experiments
- **Experiment with five or more factors:** These mainly use fractional factorial, screening studies, or Taguchi orthogonal arrays

Terms used for Design of Experiments

Factors: They are items which influence the outcome, such as temperature, time, speed, or weight. They can be either qualitative or quantitative.

Levels: They are values assigned to factors. For example, temperature levels could be set at 80 º F and 100 º F to determine which level has more impact on the outcome.

Response Variable: This is also called the dependent variable. It is the outcome observed or measured in an experiment. An experiment can have one or more response variables. Examples include door closing effort, wind noise, and water leak. Every effort should be taken to make response variables quantitative, repeatable, and accurate.

Effect: This is the change in the response variable that occurs as a factor is changed from one level to another.

Treatment Combination: These are the levels at which all factors are set, when a test run is made in an experiment.

Replication: It means repeating experiments two or more times, in order to gain greater confidence in the results. It provides insight into the magnitude of variation in the experiment due to nuisance variables.

Randomization: It involves conducting various treatment combination experiments in an order determined by chance. Experiments need to be run in a random order to remove any bias that may be introduced, as a result of that order. Randomization minimizes impact of nuisance or noise variables and increases the validity of the experiment.

Blocking: It is the arrangement of experiments into blocks, to remove known but irrelevant variations, thus improving the precision of the output.

Power: It is a measure of the sensitivity of an experiment. It shows how well the test will detect differences among levels. Power is also used to determine sample size for the experiment.

Confounding: It is the effect of one independent variable mixing up with the effect of another independent variable.

Order: It is the sequence in which the experiment is conducted. By randomizing the experiment, we reduce bias that could result from running the experiment in a logical order.

Interaction: When the effect of one input factor on the output depends upon the level of another input factor, we say that there is an interaction between these two factors. When interaction is large, the corresponding main factors have less influence.

Factorial experiments: Factorial experiments allow us to detect the interrelationships of various factors on the result. This would not be feasible in the 'one factor at a time' experimental approach.

Background Variable: It is a characteristic that potentially can affect the response variable in an experiment, but is not of interest as a factor. It is sometimes called noise variable or blocking variable. Typical background variables are batch to batch variation, time, operators, etc.

Nuisance Variable: It is a characteristic that can affect a response variable in an experiment, but is not known at the time the experiment is planned. Typical nuisance variables are environmental such as temperature or humidity, drift in measurement equipment, machine warm-up time, etc.

10 Steps in doing design of experiments

1. Determine objective
2. Identify response variable
3. Evaluate ability to measure response variable
4. Select factors and their levels
5. Select combinations and number of observations
6. Ensure data randomization
7. Run experiments and collect data
8. Analyze results
9. Verify data and results
10. Incorporate learnings into the process

Each of these actions must be carefully planned, or the value of the experiment will be reduced. The objective is to generate greater knowledge about a process. The DOE exercise should be used to reduce variation and move quality closer to target.

1. Determine objective

In order to design an experiment, a problem should be selected and defined. This will determine the setup of the experiment and actions required to be accomplished. Many experiments fail because their responses cannot be measured quantitatively. An example is visual inspections for quality judged as good or bad. Applying a scale from 1 to 5 is better than using a pass-fail method.

2. Determining response variable

The response or dependent variable is the output that is being measured in the experiment. Dependent variables can be performance measures, such as how long it took for the participant to complete the task or number of mistakes made. They could be subjective such as whether or not participants preferred the method used. Finally, they could be physical responses such as injuries that occurred. It is important to use existing knowledge or a screening method to minimize the number of factors to study.

3. Evaluate ability to measure response variable

The response variable or output should be easy to measure. This measurement should be quantitative, repeatable, and accurate. If a quantitative measure is not feasible, only then consider a qualitative one. This is usually a metric which one would want to improve.

4. Select factors and their levels

The number of levels factors have determine the number of experiments to be conducted. The levels of factors selected for study in the experiment are fixed at certain values of interest. For example, temperature may be the factor chosen and its applicable range is only from 210 º F to 230 º F. These two values could become the levels for the factor 'temperature'. If levels are too close together, the response may not be able to discriminate between them well enough, and the information obtained may not be clear. Just as for factors, levels also should be selected based on process experience and judgment.

5. Select combinations and number of observations

Select the appropriate full factorial, fractional factorial, or Taguchi orthogonal array. Choosing the appropriate matrix is important. For example, if we have too many factors and very few runs, we may lose insights of the interactions.

Another point to note is that more than one data point should be collected for each run. Obtaining only one observation may not be enough. These multiple data points are called replications. For example, if we want to study 7 factors, and all of them have two levels, the required number of runs in a full factorial experiment would be $2^7 = 128$. To study 10 factors at two levels each, we would need $2^{10} = 1,024$ runs in the experiment. Because each run may require time consuming and costly set ups, it may not be possible to perform all of them. In these scenarios, fractional factorials are used that reduce information about the interaction effects so that main effects may still be computed correctly.

6. Ensure data randomization

A designed experiment should be performed with randomized data. This will eliminate bias in the results. The order in which experiments are run should be randomized to avoid the influence of various uncontrollable factors.

7. Run experiments and collect data

Experiments should be run on a schedule and to collect data. Ensure that data is collected at a time while results are relevant.

8. Analyze results

Once data is collected, look for factors that have the largest effect on the response. Some factors may directly have a large impact on the result. These are the main effects. Other factors, when combined together may impact the response. These are called interaction effects.

9. Verify data and results

Data verification is an important part of the process. Any skewed data should be corrected and randomization methods should be relooked at. Confirmatory runs may be done to verify results. This means that a few additional experiments are run at the recommended settings to see if the desired response is achieved, and can be turned on and off.

10. Incorporate learnings into the process

Once results are collected, analyzed and validated, then it is time to make actual process changes. Change the process and monitor it with a control chart to assure that the desired response is achieved and maintained.

Developing full factorial designs

The full-factorial design tests the effects of multiple factors at the same time. For example, effects of temperature, pressure, and cycle time can be evaluated simultaneously. So, the setup of factors and levels would look like:

Factors	Levels
Temperature	100 – 150 degrees F
Pressure	5 psi – 20 psi
Cycle Time	10 min – 50 min

In a 2 level design, each factor is held at 2 predetermined values. These two levels are labeled high and low, or + and - respectively.

A full factorial experiment involves all possible combinations of factors and levels. For 3 factors, each at 2 levels, there are 2 x 2 x 2 = 8 combinations of factor settings. 2 x 2 x 2 is also written as 2^3 with the superscript 3 showing the number of levels multiplied together. A few sample layouts are shown below.

1. Two factors, two levels each 2^2 = 4 combinations

Run	Factor A	Factor B	AB	Response
1	-	-	+	
2	+	-	-	
3	-	+	-	
4	+	+	+	

2. Three factors, two levels each 2³ = 8 combinations

Run	A	B	C	AB	AC	BC	ABC	Output
1	-	-	-	+	+	+	+	
2	+	-	-	-	-	+	+	
3	-	+	-	-	+	-	+	
4	+	+	-	+	-	-	-	
5	-	-	+	+	-	-	+	
6	+	-	+	-	+	-	-	
7	-	+	+	-	-	+	-	
8	+	+	+	+	+	+	+	

The above design shows that these factors and levels present 8 different combinations. Substitute temperature for A, pressure for B, and time for C; and 100F for low, 150F for high. The first treatment combination row is for all factors at their low (-) levels; the second treatment row is for factor A (temperature) at its high level, and the other factors at their low levels.

3. Four factors, two levels each 2⁴ = 16 combinations
The number of combinations quickly increases as we increase the number of factors. Soon we reach a stage where it would not be feasible, both from a cost and time point of view, to conduct so many experiments. Hence, we introduce the idea of a fractional factorial. When compared to a full

factorial experiment, fractional factorials have fewer runs, but give up insights into the impact of the interactions.

The full factorial experiment allows us to study interactions between factors. The role of one factor, given different values of another factor can be evaluated. For example, we can test if a paint process is affected one way with high temperature and high pressure and another way if one or both of those factors is at a lower value. The 2^k formula shows the number of treatment combinations, or different experimental setups that will be generated by a two-level full factorial experiment. If two factors are studied at two levels, the number of treatment combinations is $2^2=4$. For a 3 factor experiment treatment combinations $= 2^3 = 8$

Developing half-fraction designs

In a full factorial design, information is available for all main effects and their interactions. When there are many factors, the number of interactions increases quickly. In many cases, higher-order interactions are negligible, so a lot of time could be spent on conducting experiments to understand the impact of the combination of multiple factors, but very little will be obtained results wise. There is usually a diminishing return of information on higher-order interactions. So, if a decision is made not to run higher order interactions, it will result in reduced runs, and therefore, reduced costs.

Developing fractional factorial designs

Fractional factorial designs are those experiments where only part of the full factorial design is used. Here is a two level, three factor fractional factorial design

Run	A	B	C	Output
1	-	-	+	
2	-	+	-	
3	+	-	-	
4	+	+	+	

Consider two factors A & B, at two levels each. Assume the effect of the interaction A x B is negligible. We can assign this interaction column to third factor C. That is, we will make C equivalent to A x B. So, we end up with only four runs for three factors, instead of the full eight runs in a full factorial. So, by using fractional factorial design we can reduce the number of runs.

The downside of running fractional factorials is that interaction effects are lost due to confounding. The resolution of the design describes the degree of confounding; the higher the number, the more resolution, and therefore, less confounding.

Plackett-Burman designs

When only a few runs can be done for an experiment, these designs may be used. This happens when it is too costly to have multiple runs. There will be some loss of information, and this should be kept in mind when analyzing the results. They are similar to screening designs.

Screening designs

If there are many factors in a design, then a screening design should be used. As the name suggests, it starts with a large number of factors, and ends up with a few. The factors that remain can be then examined more thoroughly. The number of runs in a screening design is usually the same as the number of factors. They are useful in the early stages of an investigation

The chart below offers some guidance on what design to select based on the objective of the experiment:

Low				High
Type of Design	Screening	Fractional Factorial	Full Factorials	Response Surface
# of factors	>4	3-15	1-7	<8
Identifies	Most important factors	Some interactions	Relationship among all factors	Optimal factor setting
Estimates	Crude direction for changes	All main effects and some interactions	All main effects and all interactions	Curvature in response

In general, consider using fractional factorials instead of full factorial designs if the impact of interactions is negligible. If the number of factors is too large, then use screening designs to eliminate some factors as candidates. Plackett-Burman designs are also be used for screening.

Calculating Impact of Factors

To find the effect of one factor:
- Sum the response values when the factor is high
- Sum the response values when the factor is low
- Subtract the low response sum from the high sum
- Divide the result by 4

To find the effect of interactions:
- Sum the response values when the factor combination is positive
- Sum the response values when combination is negative
- Subtract the low response sum from the high sum
- Divide the result by 4

These steps will make more sense when reviewed with the examples that follow.

An Example

Our goal is to find the best way to grow a tomato plant. Since we have some experience with this, we select the factors and establish levels for each factor. The factors selected are:
• Amount of water to be provided
• Type of soil to use
• Hours of sunlight seeds receive per day

When the plants have grown, we define the best plant based on its height. This is a quantitative measure, and will select the winner with very little judgment involved.

Step 1: Determine objective: The objective is to determine which factors, and their interactions influence the healthiest growth of a Tomato plant, within a given Timeframe, from its seed.

Step 2: Identify response variable: In this case we have one response variable, which is the height of the plants.

Step 3: Evaluate ability to measure response variable: In our case, the response variable is easy to measure with a scale. This measure is quantitative, repeatable, and accurate.

Step 4: Select factors and their levels: soil, water, and hours of sunlight are the factors selected. The levels are defined as follows:
soil as a two-level factor (grade A vs. grade B)
water as a two-level factor (half cup vs. 1 cup per day)
hours of sunlight is a two level factor (2 hours vs. 4 hours per day)

Define Factor Constraints
The time allowed for this experiment is five weeks, and the temperature on the thermostat is set to 70 degrees Fahrenheit.

Add Interaction Terms
The effect of any factor on the height of the plants may depend on the value of some other factor. For example, for the seed variety A, the effect of a change in water could be larger than the effect of the same change using seed variety B. This effect of factors acting together is called a two-factor interaction. These interactions are part of the full factorial design.

Step 5: Select combinations and number of observations: Here the number of factors = 3, and each factor is at two levels. If a full factorial is selected, then the number of combinations = $2^3 = 8$

Run	Soil (S)	Water (W)	Sun-light (L)	SW	SL	WL	SWL	Plant Height
1	-	-	-	+	+	+	-	
2	-	-	+	+	-	-	+	
3	-	+	-	-	-	+	+	
4	-	+	+	-	+	-	-	
5	+	-	-	-	+	-	+	
6	+	-	+	-	-	+	-	
7	+	+	-	+	-	-	-	
8	+	+	+	+	+	+	+	

Step 6: Ensure data randomization
Data randomization is required to prevent results from being skewed. In our example, we need to make sure that water is not given to the plants in the same order, and pots are not placed in the same location every day, but rotated around.

Step 7: Run experiments and collect data
Run the experiment based on a pre-determined schedule. This means that the time when water is to be given should be fixed. The other variables, soil and sunlight will remain constant.

Run	Soil (S)	Water (W)	Sun-light (L)	SW	SL	WL	SWL	Plant Height
1	-	-	-	+	+	+	-	10
2	-	-	+	+	-	-	+	4
3	-	+	-	-	-	+	+	6
4	-	+	+	-	+	-	-	2
5	+	-	-	-	+	-	+	7
6	+	-	+	-	-	+	-	6
7	+	+	-	+	-	-	-	6
8	+	+	+	+	+	+	+	3

Step 8: Analyze results
Soil effect (S):
$((7+6+6+3) - (10+4+6+2))/4 = (22-22)/4 = 0/4 = 0$

Water effect (W):
$((6+2+6+3) - (10+4+7+6))/4 = (17-27)/4 = -10/4 = -2.5$

Sunlight effect (S):
((4+2+6+3) − (10+6+7+6))/4 = (15-29)/4 = -14/4 = -3.5

S × W:
((10+4+6+3) − (6+2+7+6))/4 = (23-21)/4 = 2/4 = 0.5

S × L:
((7+6+6+3) − (10+4+6+2))/4 = (22-22)/4 = 0/4 = 0

W × L:
((10+6+6+3) − (4+2+7+6))/4 = (25-19)/4 = 6/4 = 1.5

S × W × L:
((4+6+7+3) − (10+2+6+6))/4 = (20-24)/4 = -4/4 = -1

Based on the analysis, we see that sunlight has the most significant effect on the results.

Step 9: Verify data and results
To verify data, we have to perform confirmatory runs. In this case, we would have to redo the experiment to verify the results. Since this would take a long time, it is not the recommended approach. In order to verify the data, ensure that the experimental setup was correct, and there was no cause for the results to be skewed.

Step 10: Incorporate learnings into the process
The main purpose of experimental design is to use the learnings gained to improve the process. In this example, use the combinations to grow plants to maximize their heights.

Ensure that adequate documentation is maintained so that reasons for the settings are known and shared by all interested parties.

Another Example:

Here is another example for a designed experiment. The circumstances around the experiment are not explained, but the calculations are.

Run	Temp A	Press B	Time C	AB	AC	BC	ABC	Response
1	-	-	-	+	+	+	-	25
2	+	-	-	-	-	+	+	45
3	-	+	-	-	+	-	+	20
4	+	+	-	+	-	-	-	45
5	-	-	+	+	-	-	+	25
6	+	-	+	-	+	-	-	50
7	-	+	+	-	-	+	-	20
8	+	+	+	+	+	+	+	45

A Effect:

$$\frac{(25+50+20+45) - (25+45+20+45)}{4} = \frac{140-135}{4} = \frac{5}{4} = 1.25$$

B Effect:

$$\frac{(45+45+50+45) - (25+20+25+20)}{4} = \frac{185-90}{4} = \frac{95}{4} = 23.75$$

C Effect:

$$\frac{(20+45+20+45) - (25+45+25+50)}{4} = \frac{130-145}{4} = \frac{-15}{4} = -3.75$$

A × B:

$$\frac{(25+45+25+45) - (45+20+50+20)}{4} = \frac{140-135}{4} = \frac{5}{4} = 1.25$$

A × C:

$$\frac{(25+20+50+45) - (45+45+25+20)}{4} = \frac{140-135}{4} = \frac{5}{4} = 1.25$$

B × C:

$$\frac{(25+45+20+45) - (20+45+25+50)}{4} = \frac{135-140}{4} = \frac{-5}{4} = -1.25$$

A × B × C:

$$\frac{(45+20+25+45) - (25+45+50+20)}{4} = \frac{135-140}{4} = \frac{-5}{4} = -1.25$$

The most important factors and factor combinations are those which cause the greatest deviation from zero. Factor B had the most important single-factor effect with the high level being important. Factor C followed, and here the low setting was the better level. Factor A did was not much of a factor by itself. The two and three factor interactions were relatively unimportant.

Statistical Process Control (SPC)

SPC stands for Statistical Process Control. SPC uses tools called control charts to monitor the state of a process. Processes do not always behave the same way all the time. This makes it necessary to track process performance over time. Control charts are a window into the state of a process, seen over time. A process could be either 'in control', or 'out of control'. Besides being in either of these two states, it could also show signs of problems, in spite of it being in control. These problems could manifest themselves as trends, or data points hugging the boundaries of the chart. Control charts have limits, called control lines. There are three kinds of control lines:

- upper control line (UCL)
- central line
- lower control line (LCL)

Understanding Data

There are two types of data:
1. Variable data, and
2. Attribute data

Variable data is continuous in nature. Examples include the length of a part, the time to complete a unit of work, or the weight of a sub-component.

Attribute data is discrete, or something that can be counted. Examples include number of scratches on a part, number of mispronunciations in a sentence, or the number of dents on a sheet of metal.

Attribute data can further be categorized as:
   a. Nonconformity, and
   b. Nonconforming unit

Nonconformity is an occurrence of a defect which does not conform to a specification or other inspection standard. It could be anything such as scratches, cracks, number of times service was interrupted, etc. One part or service event could have one or several nonconformities.

A nonconforming unit is a part or service event that is unacceptable because of the presence of one or more defects.

Types of Control Charts

There are two main types of control charts. They are categorized as:

a. Variable control charts
b. Attribute control charts

Variable control charts are used with measurements which are continuous in nature, such as length, circumference, time, size, weight, degrees of temperature, torques, gaps, etc. Commonly used variable control charts are the X bar and R chart, and individual and moving range chart. These charts track process, part, or service performance over time, and signal any abnormal shifts so as to quickly detect issues, and fix them to bring the process back in control.

Attribute control charts use discrete measurement points, such as number of defective parts, or the count of flaws in a part. The commonly used attribute control charts are p, np c, and u. Assumptions for p and np charts are based on the Binomial distribution, of either something is defective, or (similar to tossing a coin). The c and u charts are based on the Poisson distribution.

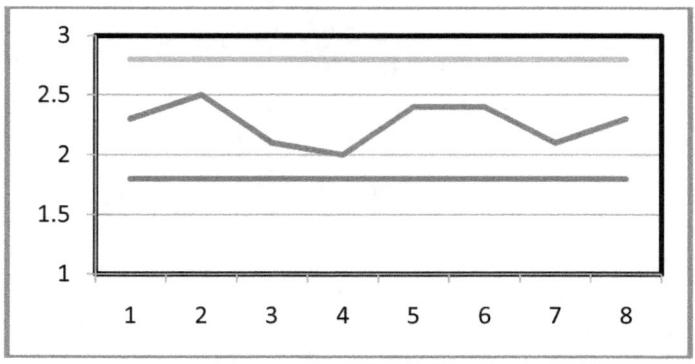

Causes of Variation

There are two causes of variation in a process. They are (a) common cause of variation, and (b) special cause of variation.

Common Cause Variation

Variation that is consistent, chronic, predictable, and inherent in a process is 'normal' or 'common cause' variation. Common causes are process variables or factors that produce the inherent variation. They are built into the process and generate a consistent overall pattern and level of inefficiency. They account for most of variation in manufacturing or in delivering a service. As they are often imperceptible to people within the process they are hard to discover and evaluate. Proper management and advanced tools are required to correct them.

Special Cause Variation

Variation that is sporadic, unpredictable, and represents a change to a process is 'special cause' or 'assignable cause' variation. Special causes factors which cause changes in the process and in the products and services those processes create. Special cause of variation is due to an external imbalanced force which destabilizes the process. They are not built into the process and are often unpredictable. Also, if left untended, they may become part of the common cause system. Special causes are easier to identify as their presence can be detected with control charts.

Function of control charts

Control charts collect and display data for a process over time. Each data point on the control chart represents a measurement for a given sample from the process.

By indicating whether processes are stable or not, control charts give clues for what should be done to improve them. After initial data collection and analysis, information on process behavior can be gained by comparing values from small-size samples to the historical pattern shown on the chart.

Control charts also help determine if changes occurring in the process output are part of random variation, or due to special causes. When a special cause change occurs, the process is deemed out- of- control, which means that something has changed. If the process behavior is unpredictable, then the source causing the instability must be determined.

Analysis of Control Charts

Control charts have a central line and control lines at the extremes. These three lines are different from specifications set for the part or process. Control lines represent the voice of the process and indicate the limits of dispersion. The central line is the central tendency of a process.

Control charts need to be analyzed to understand the performance of the process. The most common flag that a process is out-of-control is to have a point outside the control limits of the chart. When this happens, it behooves the observer to determine the special cause for this variation. Even if points are inside the control limits, there are patterns that should be kept track of, to understand how the process is performing. In addition to the point out scenario, there are four patterns which are signs that the process has changed. This means that special causes may be affecting the process. The signals are:

1. A **point out** happens when a point falls outside the control limits
2. A **run** happens when seven or more consecutive points fall on one side of the center line
3. A **trend** happens when seven or more consecutive points move up or move down
4. A **cycle** happens where data forms a repetitive pattern
5. **Hugging** happens when seven or more consecutive data points lie very close to a control limit line

These patterns are summarized as follows:

Point Out

A point plotted outside the control line is a classic signal for an unstable change in the process. Control lines are set at three standard deviations above and below the center line of the chart, so over 99% of points plotted on the graph should fall within these lines while the process is in control. It is only when a process is out-of-control should we expect to see points fall outside the control lines.

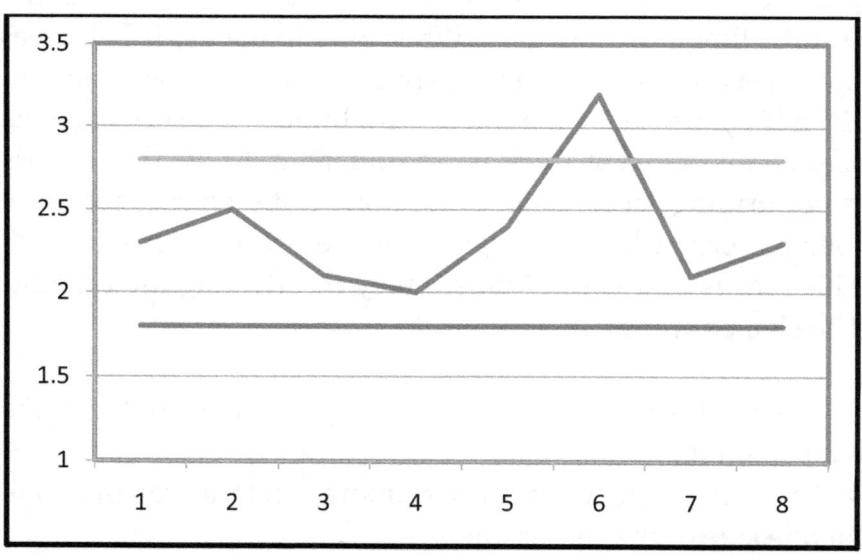

Run

A 'run' occurs when seven or more consecutive points lie on one side of the center line. When this happens, the process should be checked for any potential issues and may require an adjustment.

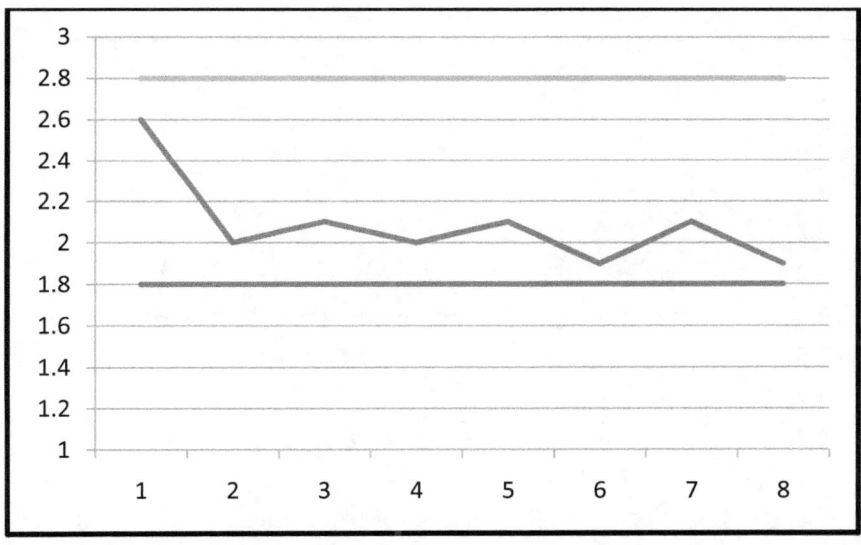

Trend

The chance of seven or more points consecutively rising or falling is similar to that of a run. It does not matter where within the control lines the trend starts or if it crosses the central line. This usually indicates that the process is slowly drifting out of control.

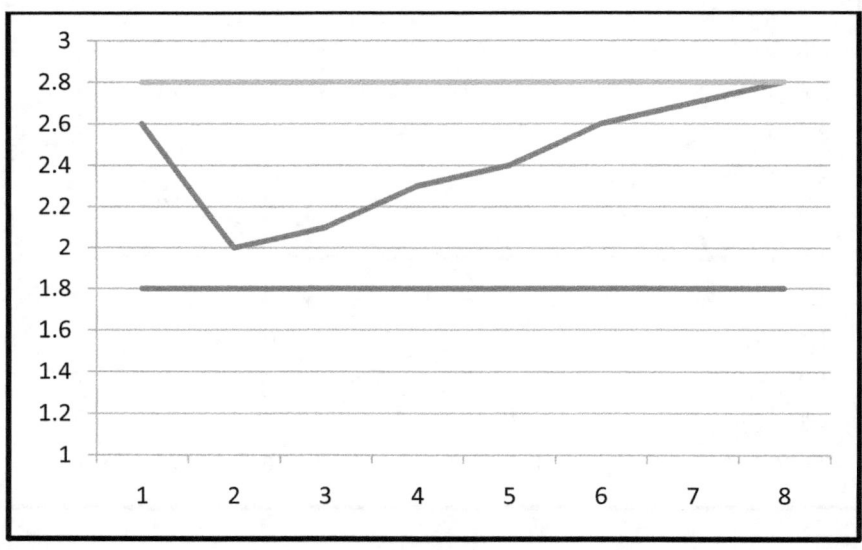

Cycle

A cycle occurs when data forms predictable repetitive patterns. When the process is in control, the position of any point on the chart should be randomly distributed. In a cycle, previous points will predict the next point's location. If the same pattern of change is seen over equal intervals, it is called 'periodicity'. This may appear as a roller coaster pattern. In this scenario, the process should be watched for something that is causing a uniform drift to both sides of the centerline.

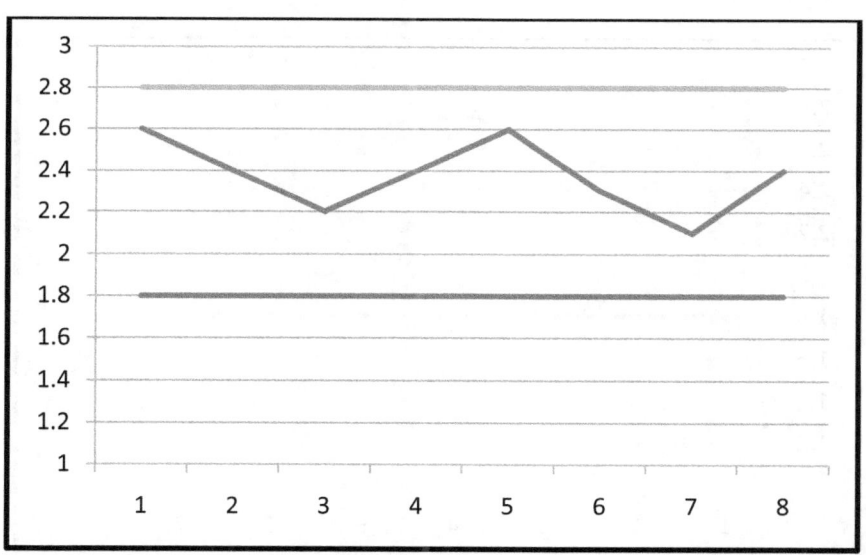

Hugging

When points on the control chart stick too close either to the center line or a control line, it is called 'hugging'. Hugging could point to some data integrity issues.

The goal is to have all points showing random variation within the control limits. This is a state of statistical stability, in which special causes of variation are absent. As process improvements are made, the points fall closer to the center line. This shows that the common causes of variation are being addressed and process variation is reduced.

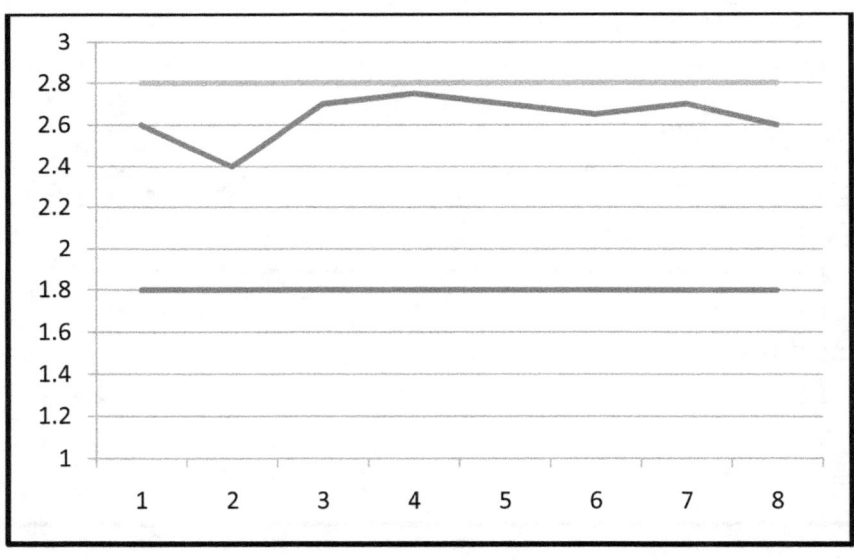

The Control Chart Process
The control charting process involves the following steps:
1. Choosing a subgroup size
2. Selecting frequency of collection
3. Deciding when to create control limits
4. Recording the raw data
5. Generating the central and limit lines

Following are the most widely used control charts for variable data:
1. X bar and R
2. Average and Range
3. Individuals and Moving Range
4. Average and Standard Deviation Charts
5. Median or Median and Range Charts

X-Bar& R Variable Control Chart

The most widely used control chart for variable data is the Xbar and R chart. The chart has two graphs, one called the average chart, and the other called the range chart. The average or Xbar chart tracks variation in the central tendency of the process. The range or R chart tracks variation within measurements by plotting the range of the largest and smallest values.

These charts are used for high-volume processes. An important part about using this chart is the formation of subgroups to collect data. For example, we could collect weld strength data to determine performance of several machines in the job shop. However, care should be taken not to mix variation from different machines. If one machine consistently has weak welds, then data from that machine will be mixed with data from the other machines, thus obscuring differences among the machines. This will not provide an accurate view of what is happening. So, it is important to form subgroups that do not mix data from various sources of variation, such as operators, machines, shifts, or fixtures. Also, mixing data over a long period of time should be avoided. In order to minimize variation sources, subgroups should be small, usually 4 to 5 data points. Data should be collected from a single source, in time order.

Example: X bar and R charts

Step 1 – Collect data from a process. This data should be collected on an ongoing basis, at regular time intervals.

Step 2 – Each data set collected represents a sub-group. For each sub-group, 4 to 5 data points should be collected. 'n' stands for number of samples in the subgroup and 'k' stands for the number of subgroups.

Step 3 - Record data on the control chart. At the end of each sub-group calculate the average and range of the sub-group The average is calculated using the following formula:

$$\bar{x} = \frac{x_1 + x_2 + x_3 + x_4 + x_5}{n}$$

The range is calculated using the following formula:

R = X (largest value) - X (smallest value)

Step 4 - Find the overall mean, or X double bar Total the mean values of Xbar, for each subgroup and divide by the number of subgroups

$$\bar{\bar{x}} = \frac{x_1 + x_2 + x_3 + x_4 + x_5}{n}$$

Step 5 – Calculate the average of the range. Total R for all the groups and divide by the number of subgroups (k).

$$\bar{R} = \frac{R_1 + R_2 + R_3 + \cdots + R_k}{k}$$

Step 6 - Calculate control limits using the following formulas. Use A2, D4, and D3 for calculating the control lines. These values are represented in a table on the following page.

Xbar control chart:
Central Line (CL) = X double bar figure you calculated
Upper Control Limit (UCL) = X double bar + A2 * R bar
Lower Control Limit (LCL) = X double bar - A2 * R bar

R Control Chart:
Central Line (CL) = R bar figure you calculated
Upper Control Limit (UCL) = D4 * R bar
Lower Control Limit (LCL) = D3 * R bar

n	A2	D4	D3
2	1.88	3.267	---
3	1.023	2.575	---
4	0.729	2.282	---
5	0.577	2.115	---
6	0.483	2.004	---
7	0.419	1.924	0.076
8	0.373	1.864	0.136
9	0.337	1.816	0.184
10	0.308	1.777	0.223

Step 7 - Construct the control chart. Draw the CL, UCL and LCL, and numerically label them. The central line is a solid line. The upper and lower control lines are usually dashed.
Step 8 - Plot the Xbar and R values for each subgroup. Note any points that lie outside the control lines.

Other variable control charts
Individual and moving range charts
I and MR charts are used for variable data. They are not as sensitive as Xbar and R charts, and this should be considered when deciding what chart to use. Individuals and moving range charts may be appropriate when:

1. The process is homogeneous
2. The process has low volume
3. It is uneconomical to take several measurements at a time

Average and standard deviation charts
These may be used if more sensitivity to process variation is desired. These charts are useful if large subgroup sample sizes are practical.

Median and range charts
These may be used when less sensitivity to process change can be balanced against ease of calculation and interpretation.

While analyzing a control chart for variable data, analyze the range chart first, since it is a good indicator of change. It is the change in dispersion which tends to affect the averages chart.

Attribute Control Charts

Attribute control charts show both, central tendency and variation on one chart. We saw that variable control charts had two charts; one to show central tendency, and another to show variation. Attribute control charts track data regarding defects, or defective items over time. The same interpretation rules of variable charts apply to attribute charts. The most commonly used attribute control charts are the c chart, u chart, p chart, and np chart.

c charts track the number of defects in a sample and are the simplest of all attribute charts. They are used where the chance of any one particular defect occurring is relatively small, but there is a good chance of different kinds of defects appearing during the collection of a sample. Since the plotted point is the total number of defects found in each sampling period, a point on the chart is easy to understand. Sample sizes are constant for c charts.

u charts track average number or rate of defects per production unit. They are commonly used in assembly or other operations scenarios where the total number of defects is collected. They relate the number of defects a production size and not to a constant sample size. This gives a defect rate relative to the amount produced.

p charts are the most sensitive of the attribute control charts and are commonly used in assembly operations where sample sizes vary. They are used to determine the proportion of defective items in an inspection unit. Since this chart uses proportions, it is possible to compare the output from inspection units of varying sizes.

np charts measure the number of defective items in an inspection unit. The sample size must remain constant from period to period of data collection. If this condition is met, and if the actual number of defective items is more meaningful or simpler to report than a proportion, the np chart may be preferred to the p chart.

Example: The np Attribute Control Chart

An np chart is used when the subgroup size is constant, while a p chart is used when it is not constant. A p chart shows the fraction defective and np chart shows number of defectives.

The np chart is used to determine if the rate of nonconforming product is stable, and will detect any deviation from stability. In some instances, only an upper control limit (UCL), and no lower control limit (LCL) is used on np charts, since rates of nonconforming product below the LCL is actually a good thing. However, by including an LCL, we could learn why low nonconforming products are made and use that knowledge to improve the process.

Steps in creating an np chart

Step 1: Collect data, consisting of the number inspected (n), and the number of defective products (np). This data should be collected over time, in sets of 50 or 100. The sample size should remain constant. Plot the sample size, and the number of defects in that sample on a chart.

Step 2: Calculate the total defects, and the total samples. Use the following formula:

$$\overline{np} = \frac{\text{number of defectives}}{\text{total number of parts inspected}} = \frac{np}{n}$$

To show a percentage, multiply by 100

Step 3: Calculate the control limits
Center line CL= $\overline{np}$

Upper control limit UCL $= \overline{np} + \sqrt[3]{\overline{np}(1-\bar{p})}$
Lower control limit LCL $= \overline{np} - \sqrt[3]{\overline{np}(1-\bar{p})}$

Step 4: Plot the lines and points on the chart to determine the state of the process.

Scenario	Chart	Calculation	Example
Attribute data for number of defects	c	$\bar{c}$ $\pm \sqrt[3]{\bar{c}}$	Number of defects in a sample of 5 laptops checked each day
Attribute data for fraction of defects	u	$\bar{u} \pm \sqrt[3]{\dfrac{\bar{u}}{a}}$	Proportion of defects based on how many sofas were manufactured each day
Attribute data for fraction of defective items	p	$\bar{p}$ $\pm \sqrt[3]{\bar{p}(1-\bar{p})}$	Fraction of requests not processed within 15 minutes
Attribute data for number of defective items	np	$\overline{np}$ $\pm \sqrt[3]{\overline{np}(1-\bar{p})}$	Requests not processed the first time through
Variable data	I MR	$\bar{X} \pm 2.66R$ $\bar{X} \pm 3.14\bar{R}$	Used to measure one measurement at a time
Variable data	Xbar R	$\bar{X} \pm A2\bar{R}$ For R charts: $UCL = D4\bar{R}$ $LCL = D3\bar{R}$	Used to measure sets of data over time

Chapter 10: Introduction

Jeff reviewed the lessons learnt from his experiences at work and decided to summarize them for easy viewing. He felt he had been through a lot, but realized that this was just the beginning. It was the introduction, not just to a new chapter, but indeed, to a new book. He felt that what he had done in the past few days was to adopt the attitude of 'learn as you go, and stumble if you have to'. He had emerged unscathed from this experience, but most wouldn't. He had had the advantage of having someone like Dean who could step in to save the day. Most wouldn't. He listed his learnings together to see if they made sense when viewed sequentially:

1. Set expectations for yourself and the team
2. Understand the technical aspects of the team's work
3. Document processes and job functions
4. Cross-train the team
5. Review results using quality tools
6. Empower
7. Summarize with statistics, action items and target dates
8. Innovate and develop new ideas and processes
9. Use statistical tools to create a structure around innovation
10. Simplify

A ten step process to get your work life in order! The steps seemed straightforward, but implementing them required a mire of people skills, subject matter knowledge, statistical and quality tools, and most importantly, the desire and ability to innovate. Consciously working towards accomplishing these steps would be the goal, and using them as a framework to set up the team's work, would be the objective. The statistical knowledge needed was to only understand the concepts behind the tools. The plan was not to be a statistician. It was to know what each tool did and use them where needed, to put some structure around innovation and development.

The final step to 'simplify' was Jeff's own addition. He had learnt a lot, and wanted to make sure that all this knowledge was not lost in complexity. There was a need to extract it from the realms of theory and implement it at the workplace. Simplification would allow that to happen. Whether it was to understand statistical tools, or management practices, the objective was to make it work in the real world. This required 'systems thinking' and the art of taking something difficult and modifying it to a point where it was easy to use.

Desired State: Surface your true potential to the top and drive innovation with the help of subject matter knowledge and six sigma quality tools.

The six sigma tools covered should be made part and parcel of the way work is done. They should not be used for the sake of using them. That is why integrating them with the subject matter will make them more potent and will facilitate the spread of innovation. The guidelines offered to Jeff were meant to bring order to the chaos that sometimes surrounds us at the workplace. Think of work as your own business. It is important to focus on the human potential by concentrating on the long term development of people and processes. Once this is done, challenge current thoughts and methods with new and different approaches.

References

1. 'Measurement Systems Analysis Reference Manual' by AIAG, Detroit, Michigan
2. 6 Sigma US http://www.6sigma.us
3. More Steam http://www.moresteam.com
4. Statsoft http://www.statsoft.com
5. Children's Mercy Hospitals and Clinics http://cmh.edu
6. Changing Minds http://changingminds.org
7. Stat Trek http://stattrek.com
8. Pathmaker http://www.skymark.com
9. 'Reliability and MTBF Overview' by Scott Speaks
10. 'Out of Crisis' by W. Edward Deming
11. 'Juran's Quality Handbook' by Dr. Joseph Juran and A. Blanton Godfrey
12. 'Quality Without Tears: The Art of Hassle-Free Management' by Philip B. Crosby

About the Author

Yusuf Biviji is an automotive and telecommunications professional with extensive industry knowhow. He holds graduate degrees in both, Engineering and Business Administration. Over the past 16 years, his experiences have spanned several areas which have shaped his views and resulted in a common-sense data driven approach to performance improvement. These thoughts are outlined in this book. If you have any comments or feedback regarding the contents of this book, you may contact him at ybiv01@yahoo.com.

Notes